Weather from Above

Janice Hill

Weather

from Above

America's Meteorological Satellites

Smithsonian Institution Press
Washington and London

Designer, Lisa Buck Vann

Library of Congress Cataloging-in-
Publication Data
Hill, Janice, 1959–
Weather from above : America's meteorologi-
cal satellites / by Janice Hill.
p.   cm.
Includes bibliographical references.
ISBN 0–87474–394–X
ISBN 0–87474–467–9 pbk.
1. Meteorological satellites.   I. Title.
TL798.M4H55   1991
551.5′028—dc20   90–30060

97 96 95 94 93 92 91 90 5 4 3 2 1

To Greg

Contents

Glossary

ABMA	Army Ballistic Missile Agency
APT	Automatic Picture Transmission
ARPA	Advanced Research Projects Agency
ASDAR	Aircraft to Satellite Data Relay
ATN	Advanced TIROS-N
ATS	Advanced Technology Satellite *later* Applications Technology Satellite
AVCS	Advanced Vidicon Camera System
AVHRR	Advanced Very High Resolution Radiometer
CDA	Command and Data Acquisition
CNES	Centre National d'Etudes Spatiales
COMSAT	Communications Satellite Corporation
COSPAS	Kosmichesaia Sistem a Poiska Avariinykh Sudov. COSPAS is the commonly used acronym in the United States; however, KOSPAS accurately represents the initials for the system. English Translation: Space system for the rescue of vehicles in distress. *See* SARSAT.
EOS	Earth Observing System
ERBE	Earth Radiation Budget Experiment
ERBS	Earth Radiation Budget Satellite
ESA	European Space Agency
ESM	Equipment Support Module
ESSA	Environmental Science Services Administration *or* Environmental Survey Satellite
FAA	Federal Aviation Administration
FGGE	First GARP Global Experiment *also known as* Global Weather Experiment
GARP	Global Atmospheric Research Program
GATE	GARP Atlantic Tropical Experiment
GMT	Greenwich Mean Time
GOES	Geostationary Operational Environmental Satellite
GOMS	Geostationary Operational Meteorological Sputnik
GSFC	Goddard Space Flight Center

HIRS   High-Resolution Infrared Radiation Sounder

HRIR   High Resolution Infrared Radiometer

HRPT   High Resolution Picture Transmission

Hugo   Highly unusual geophysical operations

ICSU   International Council of Scientific Unions

IPOMS   International Polar Orbiting Meteorological Satellite

IRIS   Infrared Interferometer Spectrometer

IRLS   Interrogation, Recording, and Location System

ITOS   Improved TIROS Operational Satellite

Lidar   Light Detection and Ranging

MCIDAS   Man-Computer Interactive Data Access System

MRIR   Medium Resolution Infrared Radiometer

MSL   Meteorological Satellite Laboratory

MSU   Microwave Sounding Unit

NASA   National Aeronautics and Space Administration

NEMS   Nimbus E Microwave Spectrometer

NESC   National Environmental Satellite Center

NESDIS   National Environmental Satellite, Data, and Information Service

NMC   National Meteorological Center

NOAA   National Oceanic and Atmospheric Administration

NOMSS   National Operational Meteorological Satellite System

NRL   Naval Research Laboratory

NSF   National Science Foundation

NSSFC   National Severe Storms Forecast Center

OMB   Office of Management and Budget

OPLE   Omega Position Location Equipment

POMS   Panel on Operational Meteorological Satellites

RAMS   Random Access Measurement System

RCA   Radio Corporation of America

RSS   Reaction Control Equipment Support Structure

SARSAT   Search and Rescue Satellite-Aided Tracking

SBUV   Solar Backscatter Ultraviolet Spectral Radiometer

SCAMS   Scanning Microwave Spectrometer

SIRS   Satellite Infrared Spectrometer

SMS   Synchronous Meteorological Satellite

SR   Scanning Radiometer

SSCC   Spin-Scan Cloud Camera

SSU   Stratospheric Sounding Unit

THIR   Temperature-Humidity Infrared Radiometer

TIROS   Television and Infrared Observation Satellite

TOS   TIROS Operational Satellite

TOVS   TIROS-N Operational Vertical Sounder

USWB   United States Weather Bureau

VAS   VISSR Atmospheric Sounder

VHRR   Very High Resolution Radiometer

VISSR   Visible and Infrared Spin-Scan Radiometer

VTPR   Vertical Temperature Profile Radiometer

WEFAX   Weather Facsimile Experiment

WMO   World Meteorological Organization

WWW   World Weather Watch

Preface

Weather from Above offers the interested layperson a brief, chronological overview of the United States's civil weather satellite efforts of the past and present. The intent is to describe in nontechnical terms the satellites that have been employed, the various instruments carried by the satellites, the data they provided, and the meteorological uses of the data. There is no attempt to probe into the political, economic, or other underlying influences upon the various programs, or to give a full indication of the complexities involved in the development of the satellite technologies. In that sense, this book is not intended to be a definitive history.

Chapter 7 describes the artifacts in the collection of the National Air and Space Museum that relate to the history of weather satellites. The satellites and instruments discussed in this section are probably the only existing unlaunched samples of certain technologies.

The bibliography combines a list of sources consulted for this book and a list of other readings that may be of interest to the general audience.

Many thanks go to Joseph N. Tatarewicz, my department chairman at the National Air and Space Museum, who gave me helpful suggestions and guidance in preparing the manuscript. I also must thank James H. Capshew and J. Gordon Vaeth for their valuable advice and suggestions.

J.L.H.

Conventional Meteorology and Unconventional Ideas

On April 1, 1960, a Thor-Able launch vehicle placed a camera-laden spacecraft, TIROS I, into orbit. That same day TIROS I sent pictures to Earth showing land masses and areas of cloud cover. The pictures were of low quality by today's standards, but they were masterpieces to meteorologists, who saw for the first time regularly transmitted images of entire weather systems.

Since then, weather satellites have evolved tremendously. The imagery has progressed from scattered, patchwork photographs taken only in daylight to highly detailed day-and-night images of the entire globe. Stepping beyond two-dimensional imagery, today's satellites give us data on the atmosphere's composition and structure. Current weather satellites also serve as communication satellites by collecting, relaying, and disseminating weather information. They have even taken on nonmeteorological roles, such as the search in search-and-rescue operations.

Before the space age, all instrument readings and observations were acquired within the atmosphere rather than above it. The earliest form of weather study occurred on a local scale. Observations were made and readings were recorded from a single location. Storms and cloud formations could be seen to pass by, and precipitation amounts could be measured. The first cloud classifications were made in the early 1800s using this method.

Information acquired from one locality, however, does not provide a large enough picture of the atmospheric conditions that will affect that locality. The atmosphere acts like intricate clockwork; each part affects the movements of all other parts. A cold air mass that enters Nebraska today may cover Kansas tomorrow, but without information from Nebraska a Kansan may misjudge how warmly to dress. Information on the entire atmosphere is needed for the best results in forecasting and other weather studies. Therefore, weather stations worldwide share their data with one another. The study and use of weather data collected from many locations is called *synoptic meteorology*. The earliest synoptic meteorologists could study past conditions only. At agreed-upon times, meteorologists in several locations would take readings and later combine their data for analysis. The invention of the telegraph allowed meteorologists to incorporate regularly disseminated data into their daily forecasts, beginning in the 1840s.[1]

One of the first supporters of telegraphic collection of weather data was Joseph Henry, the first secretary of the Smithsonian Institution. Henry arranged with sev-

Figure 1. An early TIROS image.
(Photograph courtesy of NASA)

eral telegraph offices for the relaying of weather data to be given most immediate priority in exchange for telegraph equipment provided by the Smithsonian. During 1849–50, the first year of the telegraph project, 150 weather stations participated. That number grew to 500 stations by 1860.[2] Daily, monthly, and annual synoptic maps were drawn and displayed by the Smithsonian. Unfortunately, many southern stations dropped out during the Civil War and never rejoined.[3]

In the early part of this century, a group of meteorologists based in Norway combined the local and synoptic methods and established some of the key principles of meteorology. This group, led by Vilhelm Bjerknes, created the Bergen methods for atmospheric models. The group used synoptic data from stations in Norway to identify fronts and create a model for the cyclone. Their studies led to the definition of the process of front occlusion and identification of air mass types. These models and discoveries provide a framework for modern meteorology.[4]

An important part of the local and synoptic methods and the development of atmospheric models is aerology, or the measurement and study of upper-air parameters,

accomplished by sending instruments aloft. Thermometers were sent aloft on kites and balloons, beginning in the eighteenth century. Later, various instruments were sent up in balloons to be collected afterward for readings on past atmospheric conditions. Until the 1920s, the only way to obtain fairly current upper-air data was to send a person up in a balloon or airplane to take measurements and then to return to the surface. Just as the invention of the telegraph had revolutionized synoptic meteorology, the invention of the radio revolutionized aerology. An instrument package was developed that could radio weather data from an unmanned balloon to a ground station. These radio meteorographs, now called radiosondes, could return data from 30,000 feet higher than an airplane could and were cheaper than a manned flight.[5]

By the 1950s, a worldwide network of weather stations was taking surface readings four times a day. According to international agreements, the readings were taken at 0000, 0600, 1200, and 1800 GMT, plus or minus one hour. A closely knit network of stations covered land areas, but ocean areas were sparsely covered. The North Atlantic, a critical data collection area for forecasting European weather, was covered by only ten stationary weather ships in 1949. Aerological stations were also sparsely located, and radiosonde readings were taken only twice a day because of the expense.[6]

Weather stations in the United States sent their data to the Weather Bureau generally by telegraph. Specific numbers of digits were allotted for numerical readings, such as pressure and temperature. Observed information was translated into a numerical code; for example, the amount of cloud cover was translated into eighths of the sky covered by clouds.[7] The Weather Bureau used the synoptic data to send out information, maps, and analyses to the stations via three communications networks.[8]

Photography from Rockets

A new, unconventional way to look at the weather was introduced in the late 1940s, when cameras were occasionally placed on research rockets. Beginning just after World War II, several groups from the armed forces, universities, and private industry joined together to launch rockets from the White Sands Missile Range in New Mexico. Captured German V-2 rockets were fired to study

Figure 2. A V-2 camera took this rare photograph of a thunderstorm cloud. (Photograph courtesy of United States Navy and Johns Hopkins University)

Figure 3. The Aerobee photomosaic of October 5, 1954. See page 76 for enlarged view.

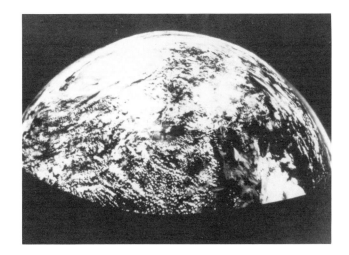

rocketry, missilery, the properties of the upper atmosphere, and solar radiation. The V-2s were followed by Viking, Aerobee, and other rocket series, launched for similar purposes.[9]

Occasionally, cameras were installed on a rocket, primarily to determine the craft's orientation at specific points during the flight. Not surprisingly, most launches of rockets carrying cameras took place on clear days, to ensure optimum launch conditions and to allow the cameras to spot ground reference points while in flight. Therefore, early rocket photographs rarely showed clouds or weather systems.[10]

The few rocket photographs that showed cloudy areas sparked interest in some meteorologists. A V-2 rocket fired on March 7, 1947, returned photos showing some cloud patterns. One meteorologist with the Air Force correlated the photos with weather conditions. He also suggested that rocket photography of atmospheric conditions had "tremendous potential usefulness," and he implied that the use of television rather than conventional photography from rockets might eliminate the problem of recovering the film.[11]

On October 5, 1954, an Aerobee rocket launched from White Sands accidentally produced the best high-altitude photograph of a weather system up to that time. Two 16-mm motion picture cameras were mounted in the nose cone of the Aerobee, perpendicular to the axis of the rocket. Near the peak of its flight, at approximately 100 miles altitude, the rocket rolled, causing the cameras to take pictures of strips of the Earth's surface. The nose section of the rocket separated and parachuted to Earth. Individual frames of the recovered film were developed and pieced together to make a photomosaic that showed a tropical storm centered over Texas. The storm had gone undetected using conventional observation and forecasting methods. The Aerobee mosaic was the first high-altitude photograph of a tropical storm.[12]

Later, rockets were launched for the specific purpose of cloud photography. Project Hugo (highly unusual geophysical operations) of the Office of Naval Research involved launching Nike-Cajun rockets out over the Atlantic from Wallops Island, Virginia.[13] But rocket photography proved to be far from an ideal system. Unpredictable rocket motion prevented the cameras from aiming at specific locations, or even simply toward the Earth. The expense was too great to be practical; each Project Hugo rocket cost seven thousand dollars in 1958, even though the camera compartment was reusable.[14] Furthermore, to be of any real use, the photographs would have to be in the hands of meteorologists within a few hours of the rocket flight, and the photographs would be needed several times a day.

Early Ideas about Weather Satellites

A few people realized that placing cameras in an orbiting vehicle might be the solution to some of the problems of complicated data collection, sparsely located weather stations, and the high cost of launching cloud cameras on rockets. As interest in space flight grew, suggestions arose for possible uses of an orbiting satellite, including weather reconnaissance.

Under contract with the Air Force, the Rand Corporation made several examinations of the possibilities for weather satellites. As early as 1946, a Rand memorandum suggested that one of the most important purposes of satellite observation of Earth might be weather observation over enemy territory. It also stated that observations of cloud patterns could be made easily and conveniently. It even proposed placing a satellite in a polar, Sun-synchronous orbit, which is exactly the type of orbit used for one entire class of current weather satellites. In 1951, Rand produced "Inquiry into the Feasibility of Weather Reconnaissance from a Satellite Vehicle," which explored the use of television on satellites.[15] Two years earlier, Eric Burgess, council member of the British Interplanetary Society, had also suggested that a television satellite's pictures would show at least the larger meteorological formations.[16]

One farsighted meteorologist of the United States Weather Bureau, Dr. Harry Wexler, not only forsaw the usefulness of a weather satellite but suggested orbital parameters and went so far as to commission a painting of what he thought the Earth would look like from the satellite, six years before the first such satellite was launched. The painting portrayed the Earth as seen at noon on the summer solstice by a satellite positioned

Figure 4. Harry Wexler's vision of how the Earth should look from a weather satellite. The small white spot near the island of Hispaniola is meant to represent a well-developed hurricane.

4,000 miles over Amarillo, Texas. Wexler had mathematically calculated the brightness of the Earth and the atmosphere, and the coloring and contrast of the sea and land areas. After surface features were painted, clouds were added. Wexler correctly predicted that a weather-observation satellite could be useful as a storm patrol, because clouds characteristic of certain types of storms could be identified and tracked. He chose common cloud patterns and storms to depict in the painting, such as frontal cyclone systems, a hurricane, a squall line, and "cloud streets," which is what he called cumulus clouds arranged in long, parallel rows. (He had noted that cloud streets were visible in the March 7, 1947, V-2 picture, along with cirrus and stratus clouds.)[17] After looking at present-day satellite photos, one can see that the painting is an amazingly accurate portrayal of the common weather systems and the general circulation patterns along the cloud streets. The only major flaw is a would-be fully developed hurricane near the West Indies depicted as a tiny spot instead of a large, spiraling feature.

In 1958, noted scientiest S. Fred Singer felt that the meteorological application of satellites would "affect our

Figure 5. This image, created by SMS 1, is somewhat comparable to Wexler's painting. (Photograph courtesy of NOAA)

way of life more than any other aspect of this space-flight business, except possibly for man to travel to the planets."[18] By that time, the projections and dreams of Wexler, Singer, and many other individuals and organizations were quickly approaching reality.

By the late 1950s, serious work on weather-observation experiments to be carried by satellites was well along.

Early prototypes had been built for what would become the TIROS Program, which would place television cameras on satellites. Also, the Army Signal Corps had begun work on the meteorological experiment for the Vanguard Program. The experiment was eventually flown on Vanguard II in February 1959, approximately one-and-one-half years after the Soviet Sputnik opened the space age.[19]

2

The First Weather Satellites

TIROS I is considered the first meteorological satellite because it was developed specifically for meteorological purposes and because of its highly successful imaging capabilities. While TIROS was still in its developmental stages in the late 1950s, three satellites carried weather-related experiments into orbit: Vanguard II, Explorer VI, and Explorer VII.

Vanguard II was part of Project Vanguard, one of the United States's contributions to the International Geophysical Year of 1957–58. Project Vanguard's specific purpose was to launch America's first artificial satellite, but the nation's first successful launch came only after the rocket team from Redstone Arsenal in Alabama had launched Explorer I. The first Vanguard satellite, a 6-inch, 4-pound test vehicle, was launched as a preliminary step toward the orbiting of a 21-pound, 20-inch spherical satellite loaded with a scientific experiment package. Several unsuccessful attempts were made to launch the 20-inch spheres, and with each attempt, a different scientific payload was housed in the sphere. Vanguard II, the craft that did reach orbit, included the cloud-cover imaging experiment created by the United States Army Signal Research and Development Laboratories. The experiment, proposed in 1956 by William Stroud and William Nordberg, would use infrared photocell units to detect the reflectivity or albedo on the Earth's surface. Albedo varies according to what type of surface is viewed. Clouds and snow-covered areas are more reflective than land areas; sea surfaces are the least reflective and, thus, the darkest. Ground personnel were to construct pictures from the albedo data. The spinning action of Vanguard II would allow the photocells to scan a strip of the Earth's surface, and the satellite's movement around the Earth would allow for scanning of subsequent lines. Unfortunately, Vanguard II wobbled, and its orbit was too elliptical. Thus no pictures could be constructed from the data.[1]

Explorer VI, launched on August 7, 1959, carried an optics system with a phototransistor to create images of the Earth. This system relayed the first television photo of the Earth, but because the craft did not move properly and its instruments malfunctioned, the lone image constructed from the data was streaked and blurry. Nevertheless, a comparison of the image with known weather conditions showed some correlations.[2]

Explorer VII was launched in October 1959, carrying a number of instruments, including a set of temperature sensors that measured the Earth–Sun heat balance. The amount of solar radiation emitted by the Sun and the

Figure 6. Vanguard II, mounted atop its launch vehicle prior to launch. In space, the satellite's antennas were extended, as shown, but for launch they were folded upward to fit inside the rocket's nose cone. An infrared photocell unit protrudes from the sphere. (Photograph courtesy of NASA)

Figure 7. The launch of Vanguard II on February 17, 1959. (Photograph courtesy of NASA)

Figure 8. The Explorer VI satellite. (Photograph courtesy of NASA)

Figure 9. A diagram of Explorer VII. The radiation detector is shown on the left. (Photograph courtesy of the United States Army)

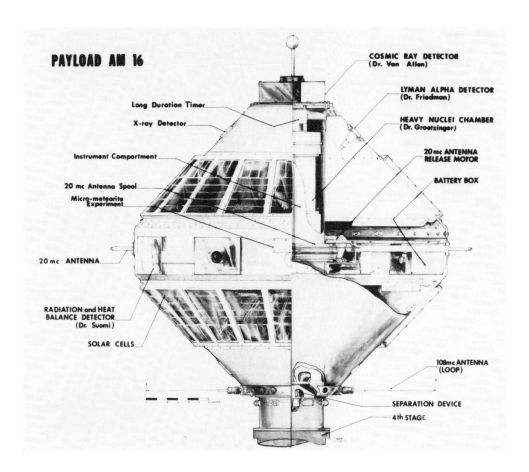

PAYLOAD AM 16

COSMIC RAY DETECTOR (Dr. Van Allen)

LYMAN ALPHA DETECTOR (Dr. Friedman)

HEAVY NUCLEI CHAMBER (Dr. Groetzinger)

20 mc ANTENNA RELEASE MOTOR

BATTERY BOX

Long Duration Timer

X-ray Detector

Instrument Compartment

20 mc Antenna Spool

Micro-meteorite Experiment

20 mc ANTENNA

RADIATION and HEAT BALANCE DETECTOR (Dr. Suomi)

SOLAR CELLS

108mc ANTENNA (LOOP)

SEPARATION DEVICE

4th STAGE

amounts absorbed, reflected, and reradiated by the Earth are very important factors in the development of weather systems. The Explorer VII sensors produced an enormous amount of useful data. Scientists were able to create maps showing geographical variations in the radiation leaving the Earth's surface. These maps matched weather maps surprisingly well.[3]

As these three satellites were developed and launched, prototypes for a television satellite were being built. Eventually these prototypes would evolve into the Television and Infrared Observation Satellite (TIROS).

The Evolution of TIROS

Between 1956 and 1960, the project that resulted in TIROS was managed by several agencies that chose several launch vehicles and defined several mission objectives for the program. Each change affected the design of satellite prototypes being developed.

The TIROS project had begun as an outgrowth of research done for an early Rand Corporation report. When the Air Force contracted Rand in the late 1940s to study and determine the feasibility of a television meteorological satellite, Rand subcontracted with RCA to discern how well television cameras could function on a spacecraft. After RCA completed its part of the Rand project, it continued to study satellite possibilities on its own.

In 1956, RCA submitted unsolicited television satellite proposals to the the armed forces and to the Weather Bureau. None of the proposals was accepted at first, but eventually RCA won a contract with the Army Ballistic Missile Agency (ABMA), at Redstone Arsenal in Alabama, to produce a television satellite that could be launched by the Jupiter C rocket.

RCA soon designed a cylindrical 20-pound prototype satellite called Janus, meant to rotate on its axis to maintain stabilization. The optimum configuration of the satellite would have been one with a large diameter-to-height ratio, that is, one that was disk shaped rather than rod shaped. Because of the Jupiter C's specifica-

tions, however, Janus was rod shaped, and engineers determined that the motions involved in launching the satellite would cause it to begin tumbling. The tumbling did occur when a Jupiter C launched the rod-shaped Explorer I, America's first satellite, in 1958.[4]

Slowing the spin rate of the satellite would also present problems. The upper stages of the Jupiter C consisted of a circular cluster of solid rocket motors developed by the Jet Propulsion Laboratory. To ensure that the thrust from the cluster was evenly distributed, and thus to keep the rocket on course, the entire unit— cluster and satellite—had to be spun at 450 RPM. A satellite carrying television cameras would need to slow that rate to between 7 and 12 RPM.[5]

In early 1958, a new RCA contract with ABMA specifically defined the program's goal, to develop a military target acquisition and location satellite. The program was now assigned the Juno II launch vehicle, which would carry a heavier, less elongated payload. To meet new design and mission requirements, RCA created Janus II, an 85-pound prototype. For the higher resolution required by the reconnaissance mission, RCA used an imaging system similar to a reflecting telescope, developed under subcontract by the Perkin-Elmer Corporation. Because the Juno II employed the solid, upper-stage cluster, Janus II had to include a device to slow the spin rate: it sent weights outward on wires to increase the total diameter. Similarly, a spinning ice skater can slow down by holding his or her arms outward.

A few months later, the Advanced Research Projects Agency (ARPA) took over management of the program. ARPA had been formed by the Department of Defense earlier that year, to manage United States space activities. ARPA assigned the program yet another launch vehicle, the Juno IV. At about the same time, the TIROS idea was conceived. The new launch vehicle allowed TIROS to be much heavier and more disk shaped than its predecessors. The plans for the Juno IV were scrapped, however, before they left the drawing board, and an Air Force rocket, the Thor-Able, was assigned to the project. Launch vehicle responsibility moved from the Army to the Air Force.

The new National Aeronautics and Space Administration (NASA) assumed direction of the TIROS project on April 13, 1959. By that time, the basic configuration of the TIROS spacecraft, including most of its subsystems, had been designed and prototypes and flight models were under construction. Still ahead lay a great deal of ground testing that would dictate critical alterations in the flight models.[6] Vibration, thermal-vacuum, acceleration, de-spin, and electrical system testing on the first flight model continued until March 20, 1960, when the craft was shipped to Cape Canaveral. TIROS I was successfully launched early on the morning of April 1.[7]

TIROS I was an eighteen-sided, drum-shaped cylinder covered in solar cells on all sides except the bottom plate. Four transmitting antennas were attached to the bottom; one receiving antenna was attached to the top. Because the Thor-Able upper stage used solid propellant, TIROS had to employ a de-spin device, a yo-yo mechanism adapted from one used on the Pioneer 4 probe. The mechanism consisted of two cables wrapped around the craft, each attached to weights. These weights would be released in flight and fly outward, until the other ends of the cables would detach from the craft. Should the craft ever rotate too slowly, small rockets mounted along the rim of the baseplate would fire to increase the rotational speed.[8]

Two television cameras mounted in the baseplate pointed parallel to the craft's spin axis and perpendicular to the baseplate. Each camera contained tiny, half-inch-diameter vidicon tubes. A tube could scan a five-hundred-line picture in two seconds, after which the picture could either be transmitted directly to a ground station within range, or stored on tape until the craft moved within range. Each camera had its own on-board recorder, transmitter, and associated electronics. One camera employed a wide-angle lens, and the other em-

Figure 12.

Figure 13.

Figures 12 and 13, this page, and figures 14 and 15, next page, show RCA technicians testing TIROS satellites. (Photographs courtesy of GE Astro-Space Division)

Figure 14.

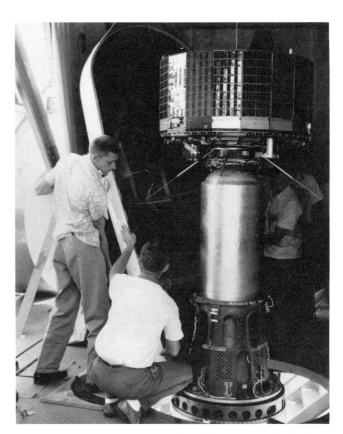

Figure 16. A TIROS satellite is in-
stalled in a Thor-Delta rocket.
(Photograph courtesy of the
United States Air Force and NASA)

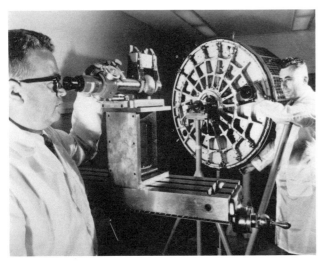

Figure 15.

Figure 17. The launch of TIROS I early on the morning of April 1, 1960. (Photograph courtesy of the United States Air Force)

Figure 18. A TIROS satellite mounted on a test pedestal. This craft, which became TIROS 4, had two wide-angle cameras protruding from its baseplate. In space, a TIROS satellite would spin about the center axis of its cylindrical body. (Photograph courtesy of NASA)

Figure 19. The dark, rectangular weight with a cable attached is the TIROS de-spin mechanism. In orbit, the weight was released and flew outward to slow the craft to an appropriate spin rate, and then the looped end of the cable would detach from the craft. (Photograph courtesy of the National Air and Space Museum)

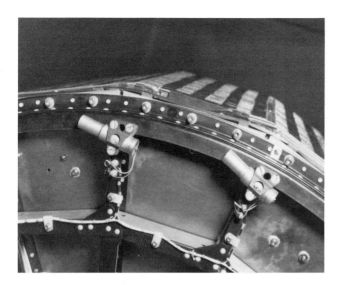

Figure 20. Two of the spin-up rockets on the base of a TIROS craft. (Photograph courtesy of the National Air and Space Museum)

Figure 21. The track of the first four orbits of TIROS I. Note how the Arctic areas were not covered. (Photograph courtesy of NASA)

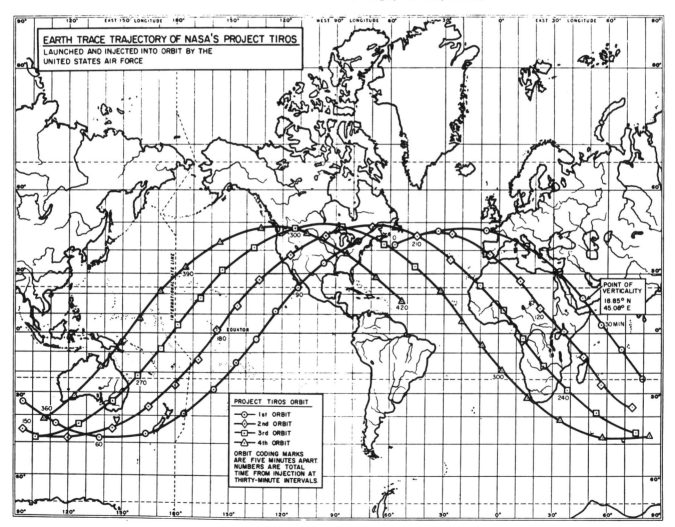

ployed a narrow-angle lens. As the TIROS I craft circled the Earth, its baseplate, and therefore its cameras, would always face the same direction. Because of this orientation, the cameras only faced the Earth for approximately half of every orbit. The orbital inclination (angle between the TIROS I orbit and the equator) was less than 50°, so polar regions were not covered. Moreover, the cameras could make pictures of the sunlit side of the Earth only, further reducing the available picture-taking time per orbit. The camera angle and the curvature of the Earth caused distortion in the resulting cloud pictures.

Nevertheless, the TIROS I pictures met with much enthusiastic praise. When President Eisenhower was shown the first pictures only seven hours after the launch, he called TIROS I a "marvelous development."[9] Senator Lyndon B. Johnson said that TIROS was the "best space news we have had in a long time . . . a tremendous step toward peace."[10] Harry Wexler, who had dreamed of weather satellites six years earlier, said, "We've suddenly gone from rags to riches overnight."[11] Not only the quality but also the quantity of pictures exceeded expectations. On April 8, one week after launch, NASA Administrator T. Keith Glennan said that the unexpected wealth of photographs required that the methods for processing and distributing them be expedited.[12]

The process for creating the pictures involved NASA's Goddard Space Flight Center (GSFC) in Greenbelt, Maryland, the Weather Bureau's Meteorological Satellite Laboratory (MSL) in Suitland, Maryland, and the Command and Data Acquisition (CDA) stations, whose locations varied throughout the program. Meteorologists at the MSL were able to select times and places for which pictures were desirable. GSFC would command the satellites when to take pictures, via the CDA stations. When satellite picture transmissions were received at a CDA station, meteorologists placed pre-prepared grid overlays on the pictures. They drew the cloud patterns seen in the pictures onto a standard, rectangular weather map and added symbols representing their own analysis of the weather conditions. These cloud maps, called "nephanalyses," were then sent by facsimile to the MSL. If the MSL felt that a "neph" was accurate, it was sent by facsimile to Weather Bureau stations, the Air Force, and the Navy. Coded nephanalyses were sent overseas by teletype. Occasionally, warnings about dangerous storms were sent immediately, before nephanalyses were drawn up.[13] The Weather Bureau once estimated that six or seven hours elapsed between the time that TIROS I created a picture

Figure 22. The first picture from a weather satellite, taken by TIROS I on the day of its launch. (Photograph courtesy of NASA)

Figure 23. An early TIROS I picture showing the Red Sea, Sinai Peninsula, and the Nile River. (Photograph courtesy of NASA)

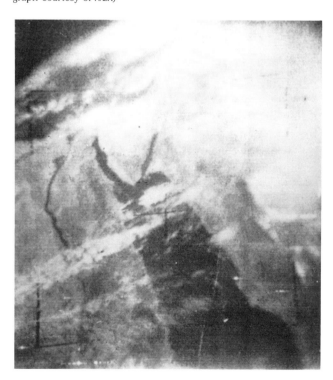

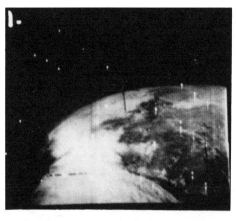

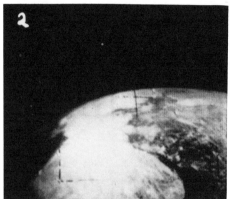
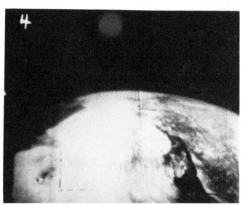

Figure 24. This sequence of TIROS I images was shown to President Eisenhower and to the press on the same day as the satellite's launch. The St. Lawrence River and the Gaspé Peninsula can be seen in the lower right corner of each. (Photograph courtesy of NASA)

and the time that meteorologists used that picture for a forecast. By the end of the TIROS program this interval was reduced to two hours.[14]

TIROS I sent pictures for three months. The CDA stations stopped interrogating the craft on June 30, 1960, after several on-board failures. Nine subsequent TIROS craft were launched through July 1965. Delta rockets launched each with no failures, a good record for the early 1960s. They orbited at altitudes of 350 to 500 miles.[15] All of the craft were similar in size and shape; weight varied slightly. Each carried the standard 500-line vidicon camera, but beginning with TIROS III, only wide-angle lenses were installed. Meteorologists found the wide-angle pictures of large-scale weather systems more valuable than the narrow-angle pictures. TIROS cameras returned a total of 649,077 pictures from April 1, 1960, to July 1, 1967.

Although the IR in TIROS stood for "infrared," TIROS I carried no infrared instruments. The United States Army Signal Corps was building the instruments at the time that NASA was formed (late 1958). Some of the key personnel working on the infrared experiments moved to NASA, but the Signal Corps would not give up the project

or the equipment. When responsibility for the TIROS project transferred to NASA in April 1959, the first TIROS flight model was too close to completion to be altered.[16]

Most of the TIROS satellites carried infrared experiments, but launches were not delayed if the IR instruments did not pass the necessary tests in time. The instruments included a scanning, five-channel radiometer, a nonscanning radiometer, and an omnidirectional radiometer. The scanning radiometer measured heat energy returned from the Earth's surface. Its detectors were mounted at an angle to the craft's spin axis to allow them to scan the area below. One channel of the scanning radiometer detected radiation in the "water-vapor window," that is, radiation that could pass through water vapor. Therefore, it detected the heat radiation directly from the Earth's surface or from cloud tops— whichever was uppermost. Meteorologists could make maps with lines of equal temperature derived from this channel's data. These maps correlated well with the cloud pictures and conventional weather maps.[17] The nonscanning radiometer provided simpler data on the total radiation from the Earth's vicinity and the total radiation from the Earth

Figure 25. A photomosaic of TIROS III pictures (right) covers most of the United States and part of the Caribbean. A nephanalysis based on this mosaic is on the left.

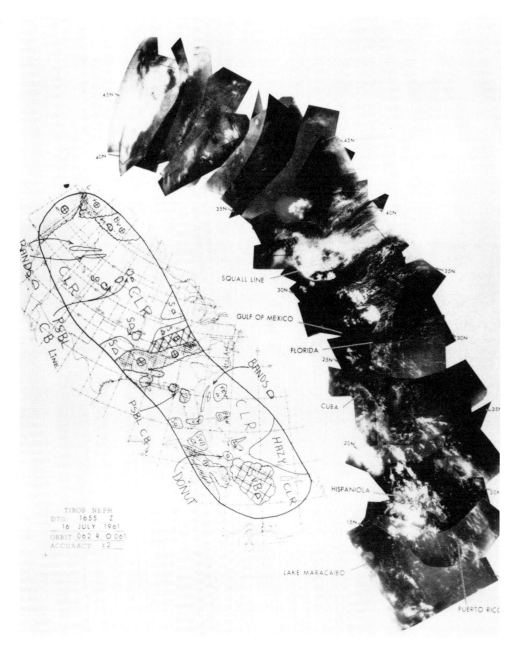

itself. The omnidirectional radiometer was similar to the radiometer carried on Explorer VII. It consisted of a black sphere and a white sphere mounted on a mirrored surface. The black sphere measured the Sun's direct radiation, the radiation reflected by the Earth, and the Earth's longwave reradiation. The white sphere measured only the latter.[18]

Weather Features Found in TIROS Pictures

During the first days after the launch of TIROS I, meteorologists had already begun learning to associate cloud types found in the pictures with weather conditions. They were surprised to see large areas covered in crescent- and doughnut-shaped clouds, which they guessed indicated areas of convection, an important method of transferring heat in the atmosphere. One of the first eas-

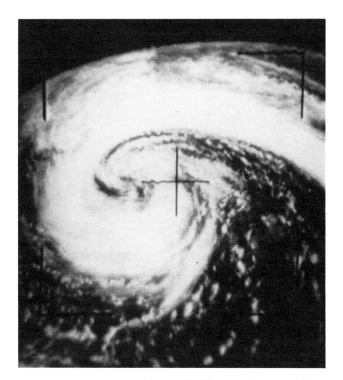

Figure 26. TIROS VII created this image of a low-pressure center and associated front on June 11, 1964. (Photograph courtesy of NASA)

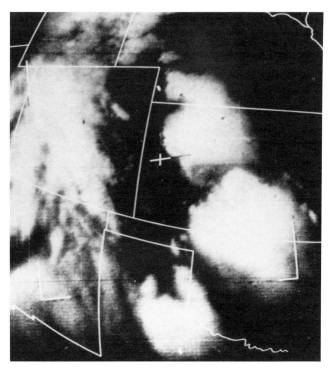

Figure 27. A TIROS image of severe thunderstorm clouds over Kansas, Oklahoma, and Texas. In weather satellite images, storm clouds are usually clearly defined, in strong contrast to land areas.

ily identifiable cloud patterns was a large circular one associated with a low-pressure center.[19] Additional cloud types were identified throughout the TIROS program. Bright, well-defined clouds were identified as areas of thunderstorm activity, and certain patterns of cirrus clouds were used to locate the jet stream.

The most dramatic features found in the TIROS pictures were spiraling clouds associated with hurricanes. Meteorologists could more easily detect and track these storms with the aid of satellite photos. TIROS III, nicknamed the "Hurricane Spy," spotted Hurricane Esther two days before conventional methods would have. The importance of early warning and the desire to learn how tropical storms develop incited NASA to launch TIROS V at the beginning of the 1962 hurricane season.[20] By 1964, meteorologists had created a system for estimating hurricane wind speeds according to four hurricane categories. Each category was defined by specific features found in a hurricane satellite picture, such as the existence and shape of the eye and the degree of organization of the clouds.[21]

One of the important innovations of the TIROS program was Automatic Picture Transmission (APT). First tried on TIROS VIII, an APT camera could scan a picture and then transmit it directly and immediately to any properly equipped user within range. An APT ground station consisted of a helical antenna, a radio receiver, and a photo-facsimile machine. Forty-seven APT ground stations participated in the original TIROS VIII test. Still in use today, APT sends data to thousands of stations, owned not only by official weather organizations but also by amateur weather buffs and a variety of educational institutions.[22]

Another important innovation of the TIROS program was the development of the "wheel configuration," first tried on TIROS IX. For this configuration, the two cameras pointed out of the satellite's sides instead of its baseplate. TIROS IX rolled along in its orbit like a wheel, and when one of the cameras pointed downward, it took a picture. The pictures showed less distortion than earlier ones because the cameras were always perpendicular to the Earth's surface when the pictures were scanned. While the wheel configuration reduced distortion, a new

Figure 28. Hurricanes Debbie and Esther appear in this mosaic of TIROS III pictures taken on September 11, 1961. (Photograph courtesy of NOAA)

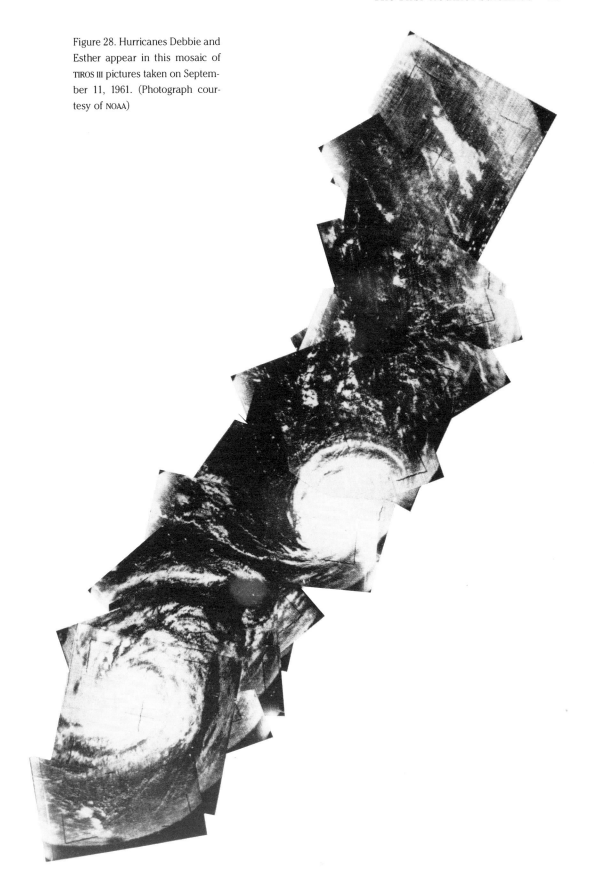

Figure 29. A helical antenna used to receive direct readout from TIROS VIII via Automatic Picture Transmission (APT). (Photograph courtesy of NASA)

Figure 30. The TIROS IX wheel configuration satellite prior to launch. A distortion calibration target is in the background. (Photograph courtesy of NASA)

type of orbit increased the amount of coverage.

TIROS IX's orbit was Sun synchronous, meaning that the craft circled over the poles, and the plane of the orbit remained aligned with the Sun year round. This orbit allowed the craft to pass over the equator at local noon and local midnight on each orbit, so that the satellite could produce long north–south strips of pictures. The strips could then be placed adjacent to each other to make a mosaic of a large area. The first mosaic of the entire Earth's weather was created from TIROS IX pictures on February 13, 1965. Because the two ends of the mosaic were taken twenty-four hours apart, the same geographical areas show different weather systems.

The TIROS program not only met its original goal of proving the feasibility of weather satellites; it also gave meteorologists valuable new views of the weather. Their appetite for TIROS data lengthened the TIROS program and strengthened the concern for continuous weather coverage. TIROS had become semi-operational, because the originally experimental satellites had become regularly used tools, providing data that meteorologists came to depend on. Meteorologists now sought a fully opera-

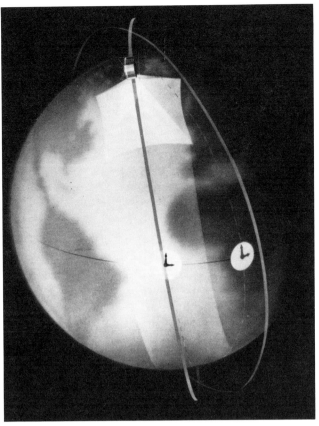

Figure 31. A wheel configuration TIROS in Sun-synchronous orbit. The satellite here would always cross the equator at 3:00 P.M., local time. (Photograph courtesy of NASA)

tional system, one consisting of satellites placed in orbit for the sole purpose of providing regular, continuous images, rather than satellites meant to test new instrumentation or orbits. Even before the first TIROS had been launched, plans for an operational system had already begun.

Development of an Operational System

As early as April 1959, NASA proposed an operational weather satellite system to Congress. The proposed system included approximately six polar-orbiting satellites called Nimbus, complemented by four geosynchronous satellites called Aeros. At that time, only two TIROS launches had been planned, so NASA suggested launching two Nimbus satellites by the end of 1961. By late 1960, Nimbus had been defined as an Earth-oriented craft with on-board stabilization, designed so that instruments could vary from flight to flight without the need to redesign the basic structure.[23]

Following the early TIROS success, the ad hoc Panel on Operational Meteorological Satellites (POMS) was formed to recommend an operational system. POMS

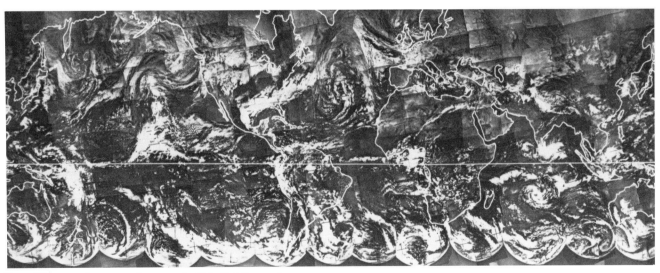

Figure 32. TIROS IX images made up this first photomosaic showing the entire Earth's weather systems. The satellite took one day to create all of the images for this mosaic. Note that the weather had changed during the twenty-four-hour period between the two orbits over Australia. (Photograph courtesy of NASA)

members included three representatives from NASA, three from the Weather Bureau, three from the Defense Department, and one from the Federal Aviation Administration (FAA). During the winter of 1960–61, POMS considered two possible operational systems: the previously proposed Nimbus, and a system based on the not-yet-tried wheel configuration TIROS. At that time, Nimbus plans had evolved further than the wheel configuration TIROS, and the proposed Nimbus was more advanced technologically and more flexible than the TIROS craft.

On April 12, 1961, POMS formally recommended that a National Operational Meteorological Satellite System (NOMSS) be developed as soon as possible. The panel further recommended that NOMSS be developed in three phases and be based on Nimbus and Aeros satellites. Standard TIROS craft would be orbited until the first Nimbus launch. NOMSS was to orbit both research-and-development and operational Nimbus satellites, with increasing operational coverage and fewer research-and-development launches as the program advanced. Aeros also would evolve into an operational system. POMS recommended that the Weather Bureau should have overall management responsibility for NOMSS, and NASA should manage the spacecraft and launch vehicle development.[24]

President Kennedy supported NOMSS and agreed that the Weather Bureau should manage it. He requested funding for the weather satellite program in his May 25, 1961, address to Congress, the same speech in which he proposed landing a man on the moon before 1970. Kennedy suggested that seventy-five million dollars go toward developing a satellite system, of which fifty-three million dollars should go to the Weather Bureau, the managing agency. A series of hearings before the House Committee on Science and Astronautics and the House Appropriations Subcommittee followed.[25] The former was concerned about whether it was more appropriate for the Weather Bureau, rather than NASA, to be responsible for a space project. The Committee on Science and Astronautics recommended that NASA and the Weather Bureau should sign an agreement defining the duties of each agency for the weather satellite project.

Such an agreement was signed in January 1962. The Weather Bureau's responsibilities included determination of meteorological requirements and use and processing of data. The responsibilities certainly did not include overall management of the program, or even design of the spacecraft. NASA was to be in charge of designing, building, and testing the craft and all operations having to with launching, tracking, programming, and communicating with the satellites.[26]

Six months after the agreement was signed, the first of several problems that would strain it occurred. NASA announced that because of delays in contract negotiations and in specification development, the first Nimbus could not be launched until mid-1963.[27] Another series of congressional hearings followed, resulting in the extention of the TIROS program to fill the satellite coverage gap.[28] Further disagreements arose over the proposed orbital altitude, the potential radiation environment of the orbiting satellite, and the locations for the CDA stations. In addition, the Weather Bureau felt that all data received from the satellite should go through the National Weather Satellite Center, whereas NASA thought that some of the data could be sent directly to users from the CDA stations.[29]

When the disagreements became insurmountable in September 1963, the Weather Bureau withdrew from the Nimbus project. As the Weather Bureau had proposed, the new operational system would be based on wheel configuration TIROS craft.[30] Nimbus became strictly a research-and-development project. A new agreement was signed by NASA and the Weather Bureau in January 1964, giving the latter much more control over the new operational system. The Weather Bureau now had the responsibility to determine all meteorological requirements, including cost and schedule, to monitor the performance of the operational system, and to manage the CDA stations.[31] NASA would no longer manage the operational weather satellite program.

3

Increased

Coverage

In the mid-1960s, three weather satellite systems were established that would define a pattern for the development of future systems. Nimbus satellites would test new instruments and thus provide data unavailable operationally. TIROS Operational System satellites, in the wheel configuration, would use proven technologies to provide data regularly. Two Applications Technology Satellites (ATS) would carry meteorological experiments to geosynchronous orbit for the first time, so that large geographical areas could be monitored almost constantly. These satellites, as well as all nonmilitary American weather satellites from that time to the present, would fall into one of the following categories: (1) Nimbus polar orbiters, used for research and development of new sensors and imaging instruments; (2) operational, polar-orbiting descendents of TIROS; or (3) geosynchronous satellites.

Nimbus

Although schedule delays contributed to the Weather Bureau's decision against using Nimbus as the operational system, the first Nimbus spacecraft was launched a year and a half before the first fully operational weather satellite. The GSFC scientists who designed Nimbus chose as its name the Latin word for "cloud." The innovative spacecraft structure consisted of: (1) a sensory ring carrying all of the Earth-observing instruments; (2) a control and stabilization system; (3) a truss connecting (1) and (2); (4) two large solar panels flanking the center section; and (5) a receiving antenna at the top of the craft.[1] Nimbus measured 12 feet in height and just over 11 feet across the panels.

NASA selected the General Electric Missile and Space Vehicle Department as the prime contractor for Nimbus in February 1961, even before the POMS had decided which system to recommend. Subcontracts were awarded to RCA Astro Electronics for cameras and the solar power systems, ITT and Santa Barbara Research Center for infrared equipment, Radiation, Inc. for telemetry, and California Computer for the command system. General Electric would build the control and stabilization system, which would eliminate the need to spin the craft.[2]

This latter system allowed Nimbus to provide for the maximum possible coverage by turning gradually through each orbit so that its instruments always pointed

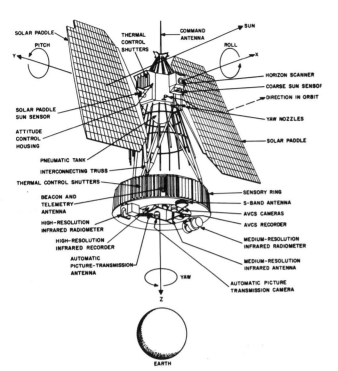

toward the Earth. As the spacecraft body turned, the two solar panels remained parallel to the Sun. Nimbus used horizon and Sun sensors to determine its attitude, and if necessary, small thrusters could correct its orientation.[3] The Earth orientation and a Sun-synchronous orbit allowed Nimbus to map the entire Earth's weather daily.

For the first time, a weather satellite had the capability of creating both day- and nighttime images. For images of the sunlit side of Earth, Nimbus carried a new camera called AVCS (Advanced Vidicon Camera System) with a 1-inch vidicon that could scan an 800-line picture.[4] Three AVCS cameras were carried on each of the first two Nimbus craft. The three cameras were mounted together, 35° apart from each other, with the center camera pointing straight down. They scanned pictures simultaneously, and because each camera took a picture of a square area 500 miles on a side, the overlapped pictures showed an area measuring 500 miles by 1,500 miles.[5] Nimbus carried a High Resolution Infrared Radiometer (HRIR). The HRIR produced images in which warm areas (land and water) were dark and cold areas (clouds) were light, similar in appearance to the cameras' pictures, but of lower resolution. The whitest areas were the coldest, therefore highest, clouds.[6]

Figure 33. A schematic of the Nimbus 2 satellite. The sensors and cameras inside the lower ring would always face Earth. (Photograph courtesy of NASA)

Figure 34. A trio of AVCS cameras. The three AVCS cameras would be mounted at angles to each other, in the Nimbus satellite's sensor ring.

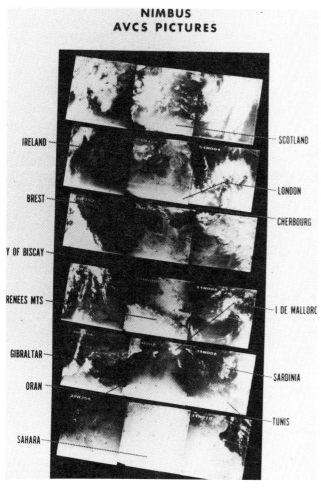

NIMBUS AVCS PICTURES

IRELAND

BREST

Y OF BISCAY

RENEES MTS

GIBRALTAR

ORAN

SAHARA

SCOTLAND

LONDON

CHERBOURG

I DE MALLORC

SARDINIA

TUNIS

Figure 35. Nimbus 1 AVCS cameras created these picture trios of western Europe and northern Af-rica. (Photograph courtesy of NASA)

Figure 36. The 85-foot-diameter dish antenna at the Gilmore Creek Command and Data Acqui-sition (CDA) station. (Photograph courtesy of NASA)

During the development of Nimbus, many redundant systems were eliminated because of Thor-Agena launch vehicle weight limitations. Nimbus underwent tests similar to the prelaunch tests of TIROS, including vibration, acceleration, and thermal-vacuum tests in the 39-foot-diameter chamber at General Electric.[7] Nimbus 1 was launched on April 28, 1964, into an orbit that was too eccentric, that reduced the time that the craft was within range of the CDA stations, and thus reduced the amount of coverage available.[8] The long-awaited satellite would be short-lived; only a month after launch, the solar panels accidentally locked into one position, greatly reducing power and making the spacecraft tumble, rendering it useless.[9] Nonetheless, Nimbus 1 returned some useful images. Nimbus 2 experienced much greater success.

Launched on May 15, 1966, it reached a nearly perfect orbit and continued to operate for two and one-half years. In addition to AVCS and APT cameras and the HRIR, Nimbus 2 carried a Medium Resolution Infrared Radiometer (MRIR) to provide information on the Earth's incoming and outgoing heat radiation. The Nimbus Technical Control Center at GSFC determined the commands given to the Nimbus satellites. CDA stations located at Rosman, North Carolina, and Gilmore Creek, Alaska, each employed 85-foot-diameter dish antennas to receive the satellite transmissions. The data were processed at GSFC and at Gilmore Creek.

Meteorologists found the Nimbus data valuable, as did other scientists. Using MRIR data, they created maps of the entire Earth showing outgoing longwave radiation, albedo, absorbed solar radiation, and net radiation flux.[10] In addition, they derived cloud-top temperatures from which cloud heights were estimated. Temperatures of sea and land surfaces were also derived, and because the Nimbus satellites regularly crossed the polar regions, temperatures of snow and ice surfaces could be monitored.[11] For the first time, satellite pictures recorded a volcanic eruption. Nimbus 2 showed a black (very hot) spot located on the newly formed island of Surtsey, near Iceland, during its volcanic eruptions of August 20 to October 3, 1966.[12]

The TIROS Operational System: ESSA Satellites

As Nimbus 2 tested sensors and systems experimentally, the new generation of TIROS satellites began providing data operationally. TOS would be managed by the new Environmental Science Services Administration (ESSA) that had been established on July 13, 1965, to control all federal environmental and geodetic activities. The Weather Bureau and the Coast and Geodetic Survey became part of ESSA, along with three new agencies: the Institutes for Environmental Research, the Environmental Data Service, and the National Environmental Satellite Center (NESC). The last agency would plan and operate environmental satellite systems, gather and analyze their data, and develop new satellite technologies to acquire environmental data. Therefore, NESC took charge of the new operational weather satellite system, including management of its CDA stations.[13]

TOS consisted of spacecraft called Environmental Survey Satellites, not coincidentally represented by the acronym ESSA. The ESSA satellites were wheel configuration TIROS craft, similar to TIROS IX. As they rolled along in their orbits, cameras mounted in their sides took pictures. Like Nimbus, the ESSA satellites utilized a Sun-synchronous orbit to provide daily global coverage, but unlike Nimbus, ESSA carried no instruments to create nighttime images. All even-numbered ESSA's carried two APT cameras. All odd-numbered ESSA's except ESSA 1 carried two AVCS cameras, like those tested by Nimbus, and radiometers similar to those on TIROS. (ESSA 1 carried standard TIROS cameras.) One even-numbered and one odd-numbered ESSA would orbit simultaneously to constitute a complete system. ESSA's 1 and 2 were launched in February 1966. Subsequent launches continued through 1969. Following the ten-out-of-ten launch suc-

Figure 37. A TIROS Operational Satellite. (Photograph courtesy of NASA)

cess of TIROS, nine ESSA's were launched successfully in nine attempts by Delta rockets. Most of the ESSA satellites orbited at an altitude of 800 to 1,000 miles.[14]

NASA had the responsibility of launching the weather satellites and monitoring them early in their flights. A few weeks after launch, NASA would transfer control to ESSA. The Gilmore Creek, Alaska, and Wallops Station, Virginia, CDA stations immediately relayed unprocessed ESSA signals to NESC in Suitland, Maryland. NESC converted the data into pictures and added computer-generated grids. Nephanalyses drawn from the pictures were sent out by facsimile. Early in the program, NESC began experimentally producing digitized cloud mosaics, high-resolution pictures of large areas. To create the mosaics, computers assigned numbers to brightness levels found in the satellite pictures.[15] By early 1967, NESC regularly transmitted these picture mosaics by facsimile to weather stations nationwide.[16]

ESSA pictures were used for the first weather satellite data exchange between the United States and the Soviet Union. Belatedly fulfilling a 1962 agreement, Washington

Figure 38. The TOS/ESSA satellites employed the wheel configuration first tried with TIROS IX. (Photograph courtesy of NASA)

Figure 39. This digital mosaic of the Northern Hemisphere was created from pictures taken by ESSA 5 on September 8, 1967. The North Pole is at the center, and North America is in the lower center portion of the mosaic. It shows several hurricanes. (Photograph courtesy of NASA)

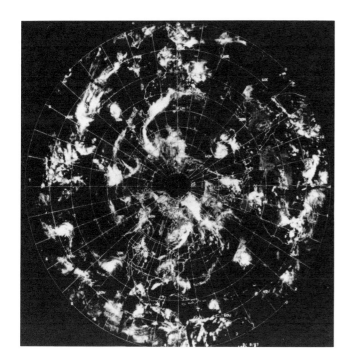

and Moscow began exchanging information in 1966.[17] At first, Moscow sent only nephanalyses. Dr. Robert M. White, administrator of ESSA, told Yevgeni Fyodorov, chief of the Soviet Hydrometeorological Service, that the nephanalyses were difficult to interpret without the accompanying satellite pictures. Soon, Moscow began sending pictures from the Cosmos 122 satellite, and Washington sent ESSA 1 pictures. Unfortunately, complicated land line routes scrambled the early Moscow picture data, but the data exchanges continued, using pictures from several Cosmos and ESSA satellites.[18]

Occasionally, ESSA satellites made headlines for providing immediate, vital information. In late 1968, heavy rains associated with Hurricane Naomi filled the basin behind a newly built Mexican dam. Officials had to decide whether to anticipate more rain and open the dam and flood a town, or leave the dam closed and take the chance that additional rain would not break the dam and flood two towns. ESSA 6 APT pictures showed that Naomi was dissipating, so the dam was left closed. The dam, the reservoir, and the two towns were preserved.[19]

In 1969, a ham radio operator received messages from the hospital ship *Hope* saying that the ship was being damaged by a heavy storm off the coast of Madagascar.

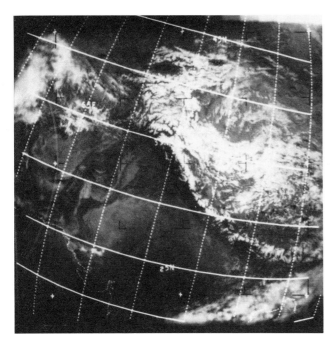

Figure 40. This ESSA 3 image shows parts of India, Pakistan, Nepal, and China. The Indus River and the snow-capped Himalayas are visible. (Photograph courtesy of NOAA)

Figure 41. This ESSA 3 image shows the weather systems over the Gulf of Mexico on October 16, 1966. (Photograph courtesy of NOAA)

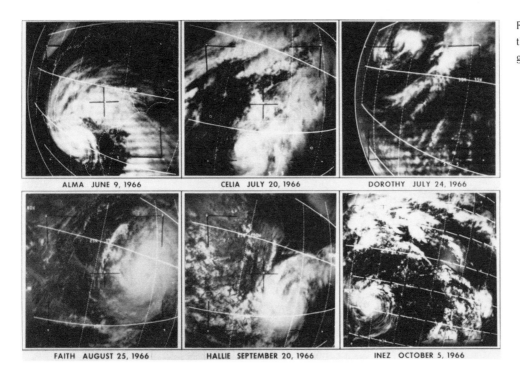

ALMA JUNE 9, 1966

CELIA JULY 20, 1966

DOROTHY JULY 24, 1966

FAITH AUGUST 25, 1966

HALLIE SEPTEMBER 20, 1966

INEZ OCTOBER 5, 1966

Figure 42. ESSA's 1 and 3 created these hurricane images. (Photograph courtesy of NOAA)

The ship seemed unable to escape the storm. The ham radio operator alerted meteorologists in Suitland, Maryland, who used ESSA 7 pictures to direct the ship safely out of the storm.[20]

The success of the ESSA program cannot be measured solely by these newsworthy events, however, nor by the sophistication of the instruments or techniques used to create pictures. ESSA simply utilized a proven technology to give meteorologists much-needed, dependable, daily coverage. System redundancy and the launching of replacement craft ensured that the flow of pictures would continue.

Applications Technology Satellites

As the ESSA satellites provided these valuable operational data and Nimbus 2 tested advanced equipment, two other satellites opened a new era in satellite meteorology. ATS 1 and 3 carried weather cameras to geosynchronous orbit for the first time.

A geosynchronous (or geostationary) satellite orbits the Earth once a day at a position over the equator. Because the satellite's orbital period is the same as the Earth's rotational period (twenty-four hours), the satellite seems to maintain the same position at all times with reference to the Earth. For example, a geosynchronous satellite placed directly above Ecuador will remain above Ecuador. The altitude required to maintain this orbit is 23,000 miles.

A geosynchronous weather satellite can make pictures of an entire hemisphere for the purpose of observing large-scale storms and weather patterns. Such a satellite provides continuous coverage of a given area rather than daily or twice-daily coverage. Continuity facilitates the tracking of storms and the observation of short-lived storms such as thunderstorms and squall lines.[21]

NASA's original 1959 proposal to Congress for an operational weather satellite system included geosynchronous satellites called Aeros, which were endorsed by POMS in 1961. Work on Aeros proceeded slowly, and when NASA realized that it could not afford the money and manpower needed to fully develop the Aeros program, the idea of a geosynchronous satellite devoted solely to weather study was temporarily shelved.

As an alternative, thought was given to using the TIROS technology to test optical equipment at very high altitudes. In late 1963, NASA began planning for an original

configuration TIROS to be launched into a extremely elliptical orbit with an apogee (point farthest from the earth) of 3,000 miles and a perigee (nearest point) of 300 miles. The laws of physics would dictate that the satellite move more slowly through its apogee than through its perigee, allowing it to take pictures from various high altitudes. This satellite would be followed by a TIROS that could reach a 22,000-mile apogee in its highly eccentric orbit, allowing it to be geosynchronous for brief periods of each orbit.[22]

NASA did not pursue either of these ideas, because the agency opted to save the time and money required to develop a new system by including a camera in a geosynchronous satellite system already under development.[23] The Syncom communication satellite and its proposed successor, Advanced Syncom, were first considered for the cloud camera project. Shortly thereafter, NASA phased out Advanced Syncom, and NASA's experiments were redirected to the new Advanced Technology Satellites, later renamed Applications Technology Satellites (ATS).[24]

NASA originally planned five ATS satellites that would test new spacecraft technologies such as stabilization systems and achievement of geosynchronous orbit. ATS would also test satellites' usefulness in communications, navigation, and meteorology. ATS 1 and 3 would be geosynchronous, spin-stabilized spacecraft. ATS 2 would orbit at approximately 6,000 miles and would employ gravity-gradient stabilization: it would send weights outward on long rods, allowing the gravitational attraction of the Earth for the weights to align the spacecraft in specific directions. ATS 4 and 5 were to test gravity-gradient stabilization in geosynchronous orbits.[25]

Unfortunately, only ATS 1 and 3 were fully successful. A launch vehicle failure sent ATS 2, which carried two AVCS cameras, tumbling in a highly elliptical orbit. ATS 4, carrying a new type of cloud sensor for day-and-night images, did not reach its appointed station.[26] ATS 5 could not attain the proper attitude in its geosynchronous orbit, but most of its planned experiments were successfully performed.[27]

ATS 1 carried a number of communications and meteorology experiments into orbit in December 1966. The cylindrical spacecraft measured approximately 5 feet high and 5 feet in diameter. Its launch weight was over 1,500 pounds, half of which was solid fuel that was expended by the apogee kick motor while placing the satellite into geosynchronous orbit.[28]

Although NASA intended ATS 1 to have a three-year life

Figure 43. The launch of ATS 1 on an Atlas Agena booster. (Photograph courtesy of NASA)

Figure 44. The ATS B (1) Spin-Scan Cloud Camera with a sample image. (Photograph courtesy of NASA)

span, it remained at least partially operable for nineteen years. During that time it performed many experiments. From its original position over the Pacific, it relayed major television events between continents and sent educational programs to remote Alaskan towns. ATS 1 also helped to save Alaskan lives. A patient suffering from an appendicitis attack was relieved by a medical aide who received a doctor's instructions via ATS 1. Similarly, a nurse was able to treat the hemorrhaging of a pregnant woman by using such a relay. The emergency treatment enabled both patients to survive until they could be transported to medical centers for further treatment.[29]

The most significant meteorological experiment was a camera to take pictures from geosynchronous orbit. NASA had originally planned to use a vidicon camera similar to those on ESSA and Nimbus for this experiment. The craft's spinning would require that an image motion compensator de-spin the vidicon pictures. But in July 1965, NASA chose a new camera for ATS that would take advantage of the satellite's spinning rather than compensate for it.[30] This ingenious Spin-Scan Cloud Camera (SSCC) could scan one horizontal line of a picture with each rotation of the craft. Because the craft spun at 100 RPM, an entire picture of two thousand lines could be

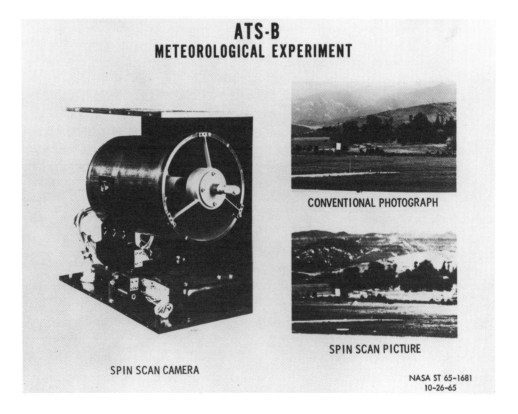

ATS-B
METEOROLOGICAL EXPERIMENT

CONVENTIONAL PHOTOGRAPH

SPIN SCAN PICTURE

SPIN SCAN CAMERA

NASA ST 65-1681
10-26-65

Figure 45. A Hughes Aircraft Company technician works on the ATS C (3) satellite. (Photograph courtesy of NASA)

Figure 46. ATS 3.

scanned in twenty minutes. The SSCC would gradually tilt with each spin (line) through a total of 15°, so the resulting picture showed almost the entire disk of the Earth. Despite its name, the SSCC was more closely related to a reflecting telescope than a camera in the strictest sense. Dr. Verner Suomi of the University of Wisconsin, who had created radiometer experiments for Explorer VII and TIROS, invented this spin-scan mechanism, variations of which would be used on geosynchronous weather satellites through the 1980s.[31]

ATS 1 carried another weather-related experiment of a different type. The Weather Facsimile Experiment (WEFAX) was designed to test the satellite's capability to relay weather data in facsimile format. NESC in Maryland sent the data, including ESSA and Nimbus satellite pictures and mosaics, to the ATS Mojave, California, ground

station. Mojave transmitted the data to ATS, which relayed it to all WEFAX receiving stations within range.[32]

ATS 3, launched in November 1967, carried experiments similar to those on ATS 1 and experienced similar success. The satellite carried a special SSCC that could create color pictures. This SSCC created the first color full-disk picture, approximately one year before Apollo 7 astronauts showed us their breathtaking views of the Earth. ATS 3 carried WEFAX instrumentation and an Image Dissector Camera that could scan pictures either vertically, or horizontally as the SSCC did. Another experiment, the Omega Position Location Equipment (OPLE), could be used for navigational as well as meteorological data collection. OPLE could locate and collect data from surface platforms or from airborne platforms such as weather balloons. From a position over Brazil, ATS 3 used

OPLE to locate within 1,000 feet a specially equipped automobile traveling at 60 miles per hour on the Baltimore –Washington Parkway in Maryland.[33] ATS 3 relayed to Europe live, color television coverage of the 1968 Mexico City Olympics and the Apollo 7 space flight.[34] It matched the longevity of ATS 1, remaining partially operable for two decades.

The ATS satellites gave us the first pictures showing our world as a round disk and featuring continents, oceans, and swirling clouds. More importantly, the satellites' stationary position allowed meteorologists to monitor the development of weather systems almost constantly. T. Theodore Fujita of the University of Chicago introduced an application of the pictures to facilitate the observation of these systems. The images were received approximately every thirty minutes, so Fujita was able to make film loops from groups of consecutive pictures. From these films, he measured cloud velocities that were found to match those calculated from radar records.[35] The cloud-velocity studies led to a breakthrough in understanding tropical circulation dynamics.[36] Movie loops became quite useful for studying short-lived developments, such as the vertical and horizontal growth of cumulus clouds.[37] NASA and ESSA established special programs to use film loops specifically to study tornado-spawning clouds. The Weather Bureau's National Severe Storms Forecast Center (NSSFC) in Kansas City, Missouri, could alert NASA if threatening weather conditions appeared. NASA would respond by command-

ing ATS 3 to scan repeatedly only the northern part of the regularly pictured area to create more pictures than usual of the critical area.[38] Through continued study, thunderstorm clouds were found to expand explosively just before tornadoes formed, when the jet stream pushed the anvil-shaped cloud tops eastward.[39]

The photographs also aided the National Hurricane Center in Miami with tropical storm prediction. By the end of the 1960s, virtually all tropical storms were discovered by satellites before conventional methods found these storms. The storms' wind speeds and intensification could be determined from satellite pictures. ESSA saved money by sending hurricane reconnaissance airplanes to previously located hurricanes only; the initial identification and location were left to the satellites.[40]

Historically, the science of meteorology made major advances when new technologies increased the amount and kind of data available. Use of the telegraph expanded the geographical area of data collection. The radiosonde allowed meteorologists to study regularly the atmosphere's vertical third dimension. Similarly, the introduction of new orbits or instruments on satellites allowed for major advances in the amount and type of data collected. In the late 1960s, Nimbus and ESSA satellites, in their Sun-synchronous polar orbits, expanded the geographical area of coverage to the maximum: the entire globe. ESSA added the necessary component of dependable service; Nimbus added the advantage of nighttime pictures. ATS added continuity of coverage of specified areas.

These satellites gave data only on the two-dimensional spherical shell of the atmosphere; they could not give data on the atmosphere's vertical dimention. Meteorologists needed to know how temperature, air pressure, water-vapor distribution, and other parameters vary with altitude in the atmosphere. Radiosondes sent aloft on balloons provided a great deal of upper-air data, from which charts were created depicting various levels of the atmosphere. Just as the radiosonde had added the third dimension to conventional meteorology, new instrumentation would add it to satellite meteorology in the 1970s.

Figure 47. The first image of the Earth from geosynchronous orbit, created by ATS 1 on December 9, 1966. See page 77 for enlarged view. (Photograph courtesy of NASA)

4

The Global, Three-Dimensional Atmosphere

The quantity and the quality of weather satellite coverage increased dramatically in the late 1960s and early 1970s. The three satellite systems established in the mid-1960s—Nimbus, TOS, and ATS—evolved into new systems with many added capabilities. Nimbus satellites carried dozens of innovative instruments to explore new ways of gathering information. Operational polar orbiters of an entirely new design carried Nimbus-tested instruments and replaced the TOS/ESSA pairs. The success of the ATS imaging systems, combined with major new efforts at international cooperation in global weather research, led to the development of geosynchronous satellites devoted solely to meteorological applications and more sophisticated methods of handling their data. These systems of the late 1960s and early 1970s represented a transitional stage between the simple picture-taking satellites and today's sophisticated sensor-carrying platforms.

Advanced Nimbus Satellites

NASA's Nimbus program continued research, development, and testing of new types of instruments on five new vehicles, Nimbus 3 through 7. The five craft were of similar design to their two predecessors—the butterfly shape with large solar panels and an Earth-oriented sensor ring to carry the instruments. The satellites were launched in 1969, 1970, 1972, 1975, and 1978 and were placed in polar, Sun-synchronous orbits at altitudes of 600 to 700 miles.[1] The instruments on each of the new satellites would be more numerous and more sophisticated than those on Nimbus 1 and 2. Some instruments would give Nimbus satellites the new capability of collecting data on the vertical structure of the atmosphere.

Nimbus B (a prelaunch designation) was scheduled to follow Nimbus 2 by carrying not only the same medium- and high-resolution radiometers, but also five new scientific instruments. Unfortunately, an improperly installed instrument in the Thorad-Agena launch vehicle sent it off course, and it had to be destroyed after lift-off.[2] A spacecraft identical to Nimbus B took its place and was orbited on April 14, 1969. That very day, the Satellite Infrared Spectrometer (SIRS) took the first sounding of the Earth's atmosphere from space. An atmospheric sounding is the measurement of temperature or other parameters at various heights in the atmosphere, over a particular location. The conventional way to take a

Figure 48. The Nimbus B space-craft prior to its attempted launch. (Photograph courtesy of NASA)

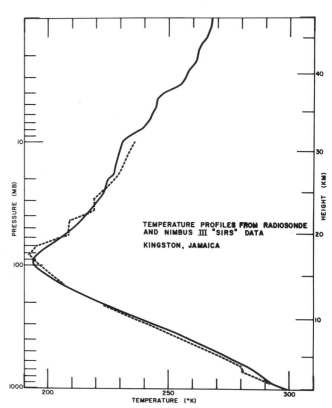

Figure 49. A graph of the first temperature sounding data from space. The solid line shows the graph of the Nimbus 3 SIRS sounding and is compared to the conventional sounding graph (dotted line) from the same time and place. Temperature decreases from the surface up to a certain altitude, then increases. (Photograph courtesy of NASA)

sounding is by sending a radiosonde up on a balloon. A satellite sounder detects electromagnetic radiation in a particular wavelength band coming from the Earth's surface. The spectrum for this radiation shows special absorption characteristics caused by conditions in the atmosphere. The SIRS collected data on the absorption of infrared radiation by atmospheric carbon dioxide. The data were used to calculate temperatures at various levels of the atmosphere. Atmospheric pressures could also be mathematically derived. The first SIRS sounding was taken over Kingston, Jamaica, where a simultaneous radiosonde sounding took place. The two resulting graphs of temperature versus height matched well. Graphs made from other pairs of readings also matched.

Yet, SIRS left room for improvement. It could not take infrared readings through clouds, and its wide field of view (100 miles) took in many areas of cloud cover, thus leaving only about 10 percent of the soundings fully use-

ful. Furthermore, SIRS could only take readings straight down, leaving large gaps in coverage between orbital paths.

SIRS was one of several sounding instruments that Nimbus satellites would test. Along with SIRS, Nimbus 3 carried another sounder, the Infrared Interferometer Spectrometer (IRIS), designed to detect not only temperature but also the water vapor and ozone content of the atmosphere. Nimbus 4 carried SIRS and IRIS instruments, improved by a narrower field of view.[3] In addition, a Filter Wedge Spectrometer detected the vertical distribution of water vapor, and a Selective Chopper Radiometer, built in the United Kingdom, profiled temperature from the cloud tops upward to an altitude of 40 miles. Sounding devices tried on Nimbus 5 included an Infrared Temperature Profile Radiometer and the Nimbus E Microwave Spectrometer (NEMS). NEMS gathered profile data in the microwave spectrum. Previous sounders detected the infrared, which gave information on areas without clouds or regions above the clouds only. Microwaves could pass through clouds, so that sounding accuracy did not depend on weather conditions.[4] Nimbus 6 carried a seventeen-channel High-Resolution Infrared Radiation Sounder (HIRS) and a microwave sounder, the Scanning Microwave Spectrometer (SCAMS).[5] Nimbus 7 carried a Stratospheric and Mesospheric Sounder, another British-built instrument, to measure vertical concentrations of water vapor and four gaseous atmospheric pollutants. Nimbus 7 monitored atmospheric ozone both vertically and spatially with the Solar Backscatter Ultraviolet/Total Ozone Mapping Spectrometer. The sounding part of this instrument measured the incident and backscattered radiation in ultraviolet wavelengths. The Nimbus satellites had, therefore, used three sections of the electromagnetic spectrum for soundings: the infrared, the microwave, and the ultraviolet.[6]

The new sounding capabilities received mixed reviews. The initial success of SIRS incited much enthusiasm from the experimenters on the project. A GSFC official was so impressed with the first SIRS soundings that he thought the new technology would lead to accurate two-week weather forecasts.[7] By October 1969, ESSA's National Meteorological Center was incorporating the sounding data in its standard operational analyses, that is, weather maps showing the state of the atmosphere at present and future times.[8] Nevertheless, scientists soon criticized SIRS and IRIS because of their dependence on a clear sky for accurate sounding down to the Earth's surface. Sounding data could be taken down to the cloud tops, but without knowing the cloud heights, the data could not be converted into useful soundings.[9] The first microwave spectrometer, NEMS, solved the problem by taking sounding data through clouds. It also successfully determined atmospheric content of water in both liquid and vapor form over the oceans. Following several tests, one study found the error of the sounding data to be within acceptable limits.[10]

In the late 1970s, several groups of meteorologists had reconsidered the usefulness of incorporating satellite sounding data into forecasting techniques. A 1977 study from the National Meteorological Center (NMC) of the National Weather Service found that data from the Nimbus 6 HIRS and SCAMS and from an operational satellite sounder had only a small impact on Northern Hemisphere forecasts, and that impact was negative. The NMC scientists found the satellite data to be of reasonable but inferior quality, and they found the data superfluous,

even over the relatively sparsely covered oceans. They concluded that the impact of sounding data on forecasts could be increased and improved by modifying the mathematical prediction model used to produce analyses and forecasts, and by improving the sounding instruments.[11]

In 1979, another group studied the effect of the same sounding instruments on forecasts. This group found that the satellite temperature sounding data could have a small positive impact on two-to-three-day forecasts. The group agreed with the NMC group that numerical prediction models could be altered to make better use of the satellite data.[12]

Figure 50. The Temperature Humidity Infrared Radiometer (THIR) on Nimbus 6 created this nighttime image of Typhoon June, located near Guam on November 19, 1975. The typhoon had winds of 150 miles per hour when the image was taken. (Photograph courtesy of NASA)

Nimbus satellites carried a number of other types of instruments along with the sounders. The Temperature-Humidity Infrared Radiometer (THIR), flown on the last four Nimbus spacecraft, provided cloud-cover maps, and by using a water-vapor channel, it provided pictures showing moist and dry areas of the atmosphere. These pictures could help meteorologists locate areas of high humidity over the oceans and tropics, where little conventional data was available.[13] The THIR was flown more as a tool to provide imaging and water-vapor data than as an experimental instrument like the sounders.[14]

Nimbus 3 and 4 used an Interrogation, Recording, and Location System (IRLS) to locate and track everything from buoys and balloons to two elk named Monique I and II. The buoys and balloons were electronically equipped to receive interrogations from IRLS, and then reply. Nimbus included a simplified version of the IRLS, the Random Access Measurement System (RAMS). RAMS did not need to interrogate its remote platforms, because they automatically initiated their own signals.[15] RAMS was used in several meteorological and oceanographic experiments, and it also tracked Japanese explorer Naomi Uemura during his dogsled trek to the North Pole in 1978.[16]

Other Nimbus instruments gave data on the Earth's heat budget, monitored solar energy, and provided Earth resources information. In addition, Nimbus 7 focused on environmental pollution by carrying several instruments that detected aerosol and gaseous atmospheric pollutants, and another that detected subtle color variations in the ocean to aid in water pollution studies.[17] Several types of Nimbus instruments, especially sounders and cloud-cover mappers, were adapted for use on operational weather satellites.

The Improved TIROS Operational Satellites

By the late 1960s, the time had come for a state-of-the-art design for the operational polar orbiters. The ESSA satellites had done their job, but they used technologies modified only slightly from that of the first TIROS. A new configuration developed by RCA Astro Electronics with NASA's direction would combine the capabilities of a pair of ESSA's into one satellite and add new capabilities that had been tested on Nimbus.

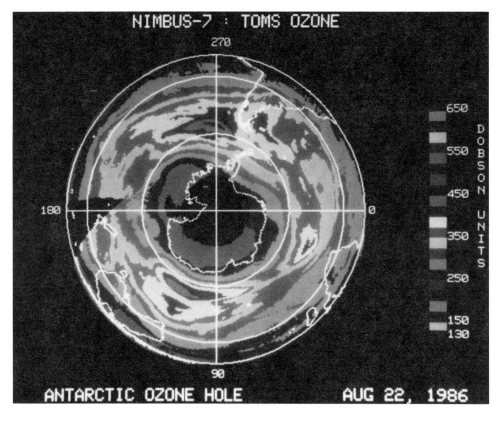

Figure 51. Images like this one have helped scientists monitor the Antarctic Ozone Hole. This image was produced at NASA's Goddard Space Flight Center from data provided by the Nimbus 7 Total Ozone Mapping Spectrometer (TOMS). In this picture, the Ozone Hole is the oval-shaped feature covering almost all of Antarctica. (Photograph courtesy of NASA)

The new satellites would be called ITOS, for Improved TIROS Operational Satellite. The central body of each was a boxy 40-by-40-by-48-inch equipment module covered in an insulation blanket. After the satellite reached orbit, three large solar panels would spring up and extend from one end of the box like three extended leaves on a drop-leaf table. The equipment module carried two AVCS cameras and two APT cameras, camera systems equivalent to those on one pair of ESSA satellites. It also carried two scanning infrared radiometers to create the first operational nighttime images. All of these instruments were mounted into one side of the module, and the satellite's three-axis stabilization made that side always face the Earth. The entire craft weighed more than 600 pounds—heftier than two ESSA's.[18] The ITOS satellites would orbit at an altitude of 900 miles and an inclination angle of close to 100° to the equator.[19]

The design of the equipment module allowed the addition of new equipment on later flights without the need to redesign the craft. The design also allowed easy access to internal components. When the thermal blanket was off, two sides could fold down into a tablelike work area.[20] After the module was folded up into the box shape and the thermal blanket was put on, a velcro-sealed port allowed limited access.

ITOS 1 was launched into a Sun-synchronous orbit on January 23, 1970, and served as an operational prototype for the new program. NASA funded the satellite but turned it over to ESSA for operational use after a five-month checkout. During the longer-than-usual checkout period, ITOS 1 exceeded all prelaunch performance requirements.[21]

Prior to launch, this first satellite had been called TIROS-M. Subsequent satellites were originally to be named ITOS before launch and ESSA 10, 11, and so on, after launch.[22] As it turned out, the postlaunch name was NOAA, after the new National Oceanic and Atmospheric Administration, which replaced ESSA and took over its operations.[23] Under NOAA, the Weather Bureau was renamed the National Weather Service, and the new National Environmental Satellite, Data, and Information Service (NESDIS) would manage the civil weather satellite systems and their CDA stations.[24] The first NOAA satellite, identical to ITOS 1, was launched in December 1970. Another identical craft, ITOS B, experienced a launch vehicle

Figure 52. The original configuration for the ITOS satellites had four cameras (two AVCS and two APT) and two Scanning Radiometers, the two protruding features near the bottom of the craft. The top sides of the three large extended panels are covered with solar cells. This ITOS is in the National Air and Space Museum collection. (Photograph courtesy of the National Air and Space Museum)

Figure 53. The TIROS M (ITOS 1 after launch) satellite, without its thermal blanket or solar panels. Abraham Schnapf of RCA is seen through the center. (Photograph courtesy of NASA)

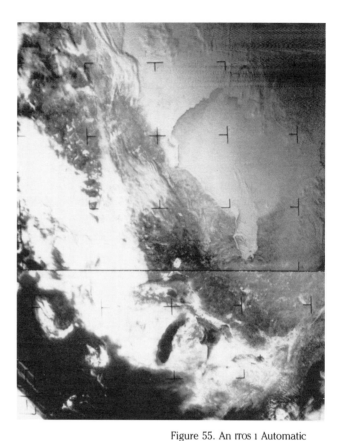

Figure 55. An ITOS 1 Automatic Picture Transmission camera created this image of the Great Lakes region and part of Canada on February 20, 1970. Lake Superior and the Hudson Bay are covered in ice and snow; Lake Michigan is clearly discernible. (Photograph courtesy of NASA)

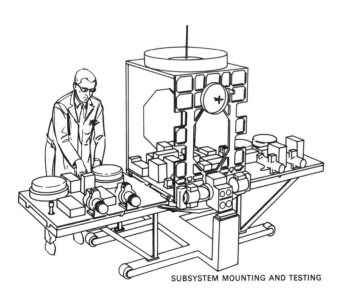

SUBSYSTEM MOUNTING AND TESTING

Figure 54. The ITOS satellites were designed so that two sides could be folded down into a tablelike work area, before the thermal blanket was in place. (Photograph courtesy of NASA)

The Global, Three-Dimensional Atmosphere 39

failure the next year.

In designing the spacecraft that would eventually become NOAA 2, RCA abandoned the use of the television camera, the device that had brought the corporation into the satellite business twenty years earlier. The new operational satellite and all of its descendants would carry radiometers for creating images. NOAA 2 was launched in October 1972 and carried a pair of scanning radiometers (SR) to create regular day-and-night images of 2-mile to 4-mile resolution. SR data could be recorded and later sent to the NMC in Washington for processing or could be sent directly to remote ground stations through APT. Some of the operational products created from SR data included pole-to-pole images, brightness mosaics, snow and ice charts, and picture mosaics of the Northern Hemisphere. The Northern Hemisphere mosaics show the Earth as if the viewer were looking straight down at the North Pole, a viewpoint that geosynchronous satellites cannot give. NOAA 2 also carried a pair of Very High Resolution Radiometers (VHRR) that created images of ½ mile resolution. NOAA used the VHRR images to create charts and then distributed these charts to show thermal features of the Gulf Stream and to support special studies. NOAA 2 provided the first operational soundings from space with its Vertical Temperature Profile Radiometer (VTPR). The VTPR had evolved from the SIRS sounder, and could sound cloudy, but not completely cloud-covered, areas.[25]

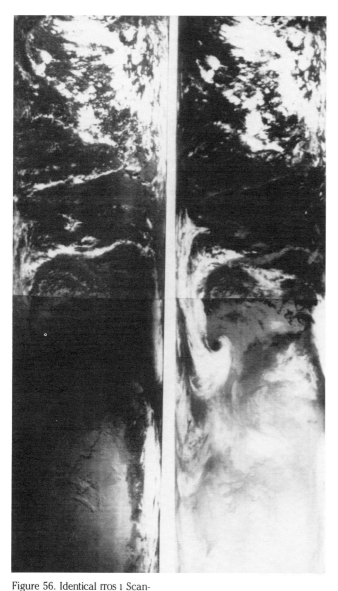

Figure 56. Identical ITOS 1 Scanning Radiometers took these images simultaneously on March 7, 1970. The image on the left, taken in visible wavelengths, shows a dark shadow created by a solar eclipse. The infrared image of the same area (right) clearly shows the cloud patterns that are shrouded in the visible image. (Photograph courtesy of NASA)

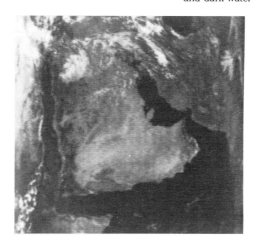

Figure 57. An ITOS 1 Scanning Radiometer created these two images of Saudi Arabia. The visible image on the left shows white clouds, light-colored land areas, and dark water areas. The infra-red image on the left shows the cooler areas in the lighter shades, and the warmer areas in dark shades. The coldest features are the clouds, which appear white. (Photograph courtesy of NASA)

Figure 58. Abraham Schnapf of RCA stands with one of the ITOS satellites of the second configuration. Shown here is the Earth-viewing side, with the apertures for the Very High Resolution Radiometers (bottom), Vertical Temperature Profile Radiometers (middle, oval-shaped), and Scanning Radiometers (top). (Photograph courtesy of NASA)

Following a second Delta launch vehicle failure in July 1973, NOAA 3 was launched in November of that year. NOAA 3, NOAA 4 (launched in 1974), and NOAA 5 (1976) were virtually identical to NOAA 2. Beginning with the launch of NOAA 3, two NOAA satellites would orbit simultaneously, to provide twice the data twice as often and to ensure that a failure would not cause a gap in coverage. The NOAA satellites continued the important job begun by the ESSA's, of bringing useful data to the NMC and independent users on a regular basis. The new polar orbiters became tools of quantitative value to meteorologists by providing data that could be plugged into the mathematical equations of their prediction and analysis models. The low orbits of the polar orbiters made them the preferred platforms for soundings and the creation of high-resolution images necessary for some areas of study. On the other hand, geosynchronous satellites were preferred for those disciplines requiring large-scale visual data at quick, regular intervals.[26]

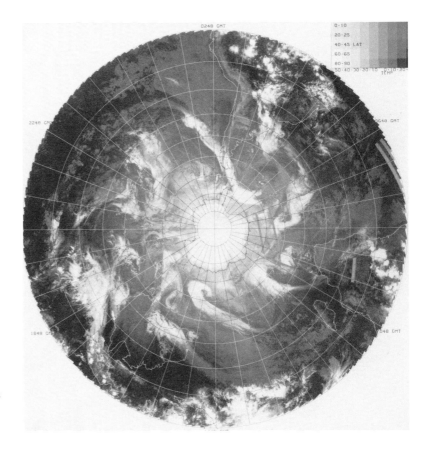

Figure 59. The NOAA 2 Scanning Radiometer created the infrared images that make up this Southern Hemisphere mosaic. Fronts appear to radiate outward from Antarctica. Geosynchronous satellites cannot give meteorologists this polar point of view. (Photograph courtesy of NOAA)

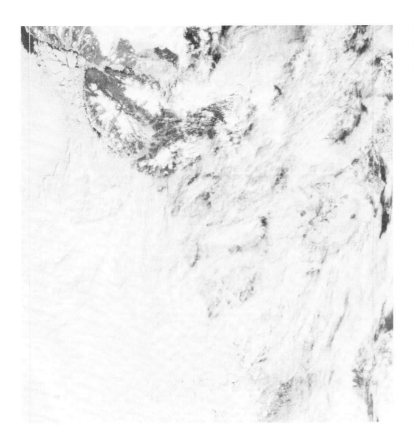

Figure 60. NOAA 4 created this visible image of the Arctic. (Photograph courtesy of NOAA)

Figure 61. An infrared image of the Arctic made by NOAA 4 showing the same region as figure 60. (Photograph courtesy of NOAA)

Figure 62. The Alaskan coast
appears in this visible
image, taken by NOAA 4.
(Photograph courtesy of NOAA)

Figure 63. This NOAA 4 infrared
picture of the Alaskan coast covers the same area as figure 62.
(Photograph courtesy of NOAA)

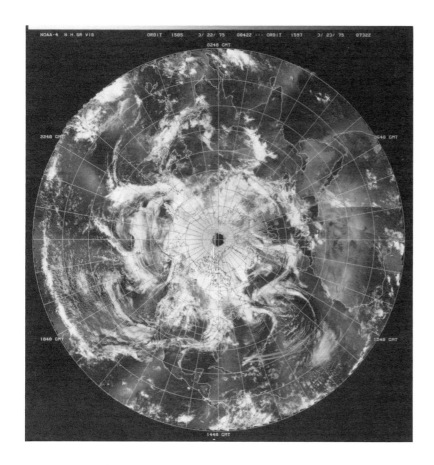

Figure 64. NOAA 4's Scanning Radiometer provided the visible-wavelength images that make up this mosaic of the Northern Hemisphere. (Photograph courtesy of NOAA)

Geosynchronous Weather Satellites

ATS 1 and 3 continued to provide the necessary geosynchronous data dependably through the late 1960s. The cloud cameras had originally been placed on the ATS's only as quick and less expensive ways of getting cameras to geosynchronous orbit. In the mid-1970s, another ATS would carry a cloud-imaging instrument to geosynchronous orbit, this time for purely experimental purposes. ATS 6 carried approximately twenty communications and technology experiments, including a VHRR specially built to create images from a geosynchronous orbit. ATS 6 had on-board stabilization, so it did not spin. The VHRR therefore created its visible and infrared images without using the spin-scan technique.[27] ATS 6 remained at least partially operable until 1979. Perhaps the most notable capability of ATS 6 was its transmission of signals powerful enough to be received by relatively small, inexpensive ground stations. This capability allowed the satellite to send health and educational programs to remote areas of India and the United States. Soon NASA and NOAA would initiate a new series of geosynchronous satellites that, unlike ATS, would be devoted solely to meteorological applications. They would continue the storm patrol begun by the ATS, and they would contribute to a Global Atmospheric Research Program (GARP).

Because the atmosphere is a single entity covering the globe, a logical way to proceed in studying it would be through global-scale weather studies achieved through international cooperation. Weather satellites are logical tools to use for global study, because they provide quick information on large portions of the atmosphere. Confident of the capabilities and potential of weather satellites, the United Nations had made recommendations for international weather study programs, beginning in the early 1960s. The first recommendation, part of United Nations Resolution 1721, was based on a National Academy of Sciences report to President Kennedy. The United Nations Resolution requested that the United Nations's World Meteorological Organization (WMO) work with various organizations, including the International Council of Scientific Unions (ICSU), to report on methods of improving atmospheric science and weather forecasting "in the light of developments in outer space." In response, WMO endorsed a World Weather Watch (WWW) to provide better coverage and suggested areas of meteorological research worthy of global study. In a later resolution, the General Assembly asked the WMO to continue its studies on global data coverage and research, and it invited the ICSU to develop an atmospheric research program to complement the WMO efforts. The latter program evolved into GARP. In October 1967, the WMO and the ICSU formally agreed to sponsor GARP jointly.[28] The satellites to be launched in the coming years would play an important role in GARP experiments.

The original operational weather satellite system that NASA had proposed to Congress in 1959 had included geosynchronous satellites stationed around the globe; now GARP would promote the creation of a ring of five such satellites that would be established through international cooperation. Those satellites first proposed in 1959 eventually acquired the name SMS, for Synchronous

Meteorological Satellite. NASA had not forgotten the name and used it for its two prototype spacecraft of the new program.

In 1970, Philco-Ford won a contract to build three satellites based on ATS 1 and 3 design. The first two were funded by NASA—their SMS craft; the third was funded by NOAA, and would be called GOES, for Geostationary Operational Environmental Satellite.[29] Later, a follow-on contract would call for two more GOES.[30] The five craft were almost identical. Each was cylindrical with a magnetometer and antennas extending from one end; overall height was approximately 11 feet and maximum diameter was just over 6 feet.[31]

SMS and GOES employed a spin-scan imaging technique similar to that used on ATS 1 and 3, for which the satellite's spinning action allowed it to scan lines of a picture. The new satellites carried an improved version of the spin-scan camera, called the Visible and Infrared Spin-Scan Radiometer (VISSR). The VISSR could scan black-and-white images in both the visible and infrared bands, providing day-and-night coverage. The VISSR had eight identical visible channels, so that eight lines could be scanned at once.[32] One picture could be created by 1,821 spins in 18.2 minutes. Therefore, a visible picture was made up of eight sets of 1,821 lines, or 14,568 lines, and had a resolution of less than half a mile. An infrared channel scanned a 1,821-line picture with a resolution of just over 5 miles. Like ATS, SMS and GOES could scan a smaller area of the Earth repeatedly at short intervals, if a particular storm system needed to be watched closely.[33]

SMS and GOES served also as communications satellites for weather data. For example, they would send their own VISSR data to NOAA's Wallops, Virginia, CDA station, where it would be processed to simplify transmission. The CDA station would then send it back to the satellite, which would change its frequency, amplify it, and send it to regional data utilization stations. Similarly, the satellites could make the same alterations to collected WEFAX data and relay it to WEFAX receiving stations. They were also able to relay data between CDA stations and automatic data collection platforms mounted on land, ships, aircraft, and buoys.[34]

Each of the satellites was first launched into an extremely elliptical orbit with an apogee close to geosynchronous altitude. The craft's on-board apogee boost motor then fired to raise the perigee. Once the satellite was in geosynchronous orbit, it would be stationed over the equator at a particular longitude. The satellite would

Figure 65. SMS 2, prior to launch. The aperture for the Visible and Infrared Spin-Scan Radiometer (VISSR) is in the center. (Photograph courtesy of NASA)

Figure 66. The logo of the GARP   shows the geographical area of
Atlantic Tropical Experiment   study for the experiment.

have to remain over the equator, but could be moved eastward or westward.

SMS 1 began its mission in the summer of 1974 at 45° west, so that it could assist in the GARP Atlantic Tropical Experiment (GATE). The position put SMS 1 directly above a point just off the north coast of Brazil. Its assignment to GATE began even before NASA had a chance to check out the satellite.[35] In fact, the experiment had been moved from the Pacific to the Atlantic during the early planning stages, partly because no geosynchronous satellite would be stationed over the Pacific at the proper time.[36]

GATE involved not only SMS 1, but also ATS 3, NOAA 2 and 3, Nimbus 5, defense satellites, Soviet Meteor satellites, and a number of aircraft and ships. NOAA 2 and 3 sent APT images to the Dakar, Senegal, base for field operations, and Nimbus 5 provided sounding data. SMS 1 images were used to direct aircraft to particular areas for study.[37] During GATE, SMS 1 became the first satellite to monitor a hurricane almost continuously during both day and night. The director of NOAA's Hurricane Center in Miami said that SMS 1's images had a "most dramatic" impact on the Center's understanding of hurricanes and other weather patterns in the Atlantic.[38]

Following GATE, NASA moved SMS 1 to its regular position at 75° west. From this point, directly above the border between Columbia and Peru, it monitored the eastern half of the United States and the nearby Atlantic and Caribbean areas where tropical storms occur. SMS 1 therefore replaced ATS 3 for weather pictures of that region, so NASA turned off the ATS 3 camera and reserved it as a backup.[39] In February 1975, the other prototype, SMS 2, was launched. After being stationed at 115° west for ten months, it was moved to its permanent station over the eastern Pacific Ocean at 135° west. From there it monitored the weather of the western United States and storms in the Pacific.[40] From this time on, NOAA would try to maintain one eastern (75° west) and one western (135° west) satellite, replacing satellites when necessary, to ensure continuous coverage of the United States and its coastal areas. GOES 1 replaced SMS 1 as the eastern satellite in late 1975, so the latter was moved to 105° west and put on stand-by status.[41] GOES 2 became the new GOES-EAST in mid-1977,[42] and one year later GOES 3 became the new GOES-WEST.[43] GOES 1, 2, and 3 participated in GARP's first global experiment from 1978 to 1979.

Any one of these geosynchronous satellites could produce an enormous amount of data. For example, SMS 1 produced approximately fifty pictures per day; each consisted of 1.5 billion data bits, so one day produces 75 billion bits. To be fully useful, this great amount of weather data must be processed quickly and efficiently. One of the less efficient ways of using geosynchronous satellite data is to convert it to a photographic format, because the excellent geometric fidelity of the spin-scan camera data is lost. In 1972, staff members of the Space Science and Engineering Center of the University of Wisconsin set out to improve methods for handling data by creating a new computer system.

They combined specially designed and off-the-shelf hardware and software to create MCIDAS, the Man-Computer Interactive Data Access System.[44] The key word is *interactive*. The operator could manipulate the data in a number of ways, by enhancing or magnifying images, combining them into loops of any length, running the loops at a variety of speeds, tracking clouds, and superimposing velocity vector plots of the cloud tracks. The operator could also run quick comparisons and interlace visible and infrared pictures. Measurements made on MCIDAS are accomplished through use of the original digital data, not photographic displays. One NASA-sponsored study of GATE data found that the

Figure 67. Geostationary Opera-   aboard a Delta rocket being pre-
tional Environmental Satellite A   pared for launch. (Photograph
(1) inside its payload fairing   courtesy of NASA)

Figure 68. Hurricane Allen, in the   this SMS 2 picture taken August 8,
Gulf of Mexico, and Hurricane   1980. (Photograph courtesy of
Isis in the Pacific are visible in   NASA)

MCIDAS method for tracking was at least as accurate as conventional methods.[45] Although the ATS and upcoming SMS pictures first inspired the creation of MCIDAS, the system was also used to analyze data from other types of spacecraft, such as a Landsat Earth resources satellite and Mariner 10, a space probe that flew by Mercury and Venus.[46]

Scientists used SMS and GOES imagery in a variety of studies. Pollution could be detected by observing that part of the atmosphere that appeared in cross section at the edges of the pictures. In these areas, scientists could observe the effects of pollutants on light passing through the atmosphere and, thus, learn about the pollutants themselves. To observe the atmosphere at the low angles required for these studies, the eastern satellite would be used for the western United States and vice-versa.[47] Large-scale spatial areas of pollution, such as areas of smog or sulfate aerosols, could be monitored

and tracked by using MCIDAS.[48]

A study cosponsored by NASA and the Institute of Food and Agricultural Sciences found GOES infrared data very useful for predicting freezes in Florida, vital information for citrus fruit growers. GOES was particularly well suited for the freeze study because of its image-making frequency and the fact that most freezes occur on cloudless nights, when the satellite could see down to the surface.[49]

GARP would make the largest-scale use of the geosynchronous satellites in its upcoming First GARP Global Experiment, also known as the Global Weather Experiment, to take place from December 1, 1978, to November 30, 1979. This first global-scale weather experiment would employ nine satellites and scores of ships and aircraft, and would make use of thousands of human- and machine-made observations from ground, ocean, and upper-air stations.[50] The Global Weather Experiment

Figure 69. GOES 1 created this image while it was stationed over the Indian Ocean for the Global Weather Experiment. (Photograph courtesy of NOAA)

would require that five geosynchronous weather satellites be stationed around the world. Experiment organizers originally planned that the five satellites would include a GOES-EAST and a GOES-WEST from the United States, one Japanese satellite, one European Space Agency (ESA) satellite, and one Geostationary Operational Meteorological Sputnik (GOMS) from the Soviet Union. Japan launched its Geostationary Meteorological Satellite (also called *Himawari*, Japanese for "sunflower") in June 1977. ESA followed in November with the launch of Meteosat. Unfortunately, the USSR announced that it would not be able to launch a GOMS in time for the experiment, so in late 1977, the WMO asked the United States and ESA to look into ways of filling the gap over the Indian Ocean. They decided to move GOES 1 to the 60° east location after GOES 3 had replaced it at 135° west.[51]

An advanced series of polar orbiters would also participate in the Global Weather Experiment and would succeed the ITOS satellites. Although no new Nimbus satellites would be launched, the latter craft in the series would continue to supply data well into the 1980s, and technologies tested by Nimbus would be applied operationally by NOAA satellites.

Of course, the addition of sophisticated capabilities to the satellites placed a high price tag on their fabrication and launching. Several failures of the satellites would strain NOAA's budget. During the all-out effort to decrease federal spending in the 1980s, possible methods for reducing the cost of the satellite program were explored, including a proposal to sell the weather satellites to private industry.

5

Modern Systems,
Modern Problems

In the late 1970s, an operational polar orbiter more sophisticated than ITOS was introduced for the global experiment and for continued but improved data collection. Geosynchronous satellites were considered as necessary as the polar orbiters, so more were launched to keep coverage continuous over the United States. As the satellites became more sophisticated, so did the problems associated with them. Both the polar orbiters and the geosynchronous satellites experienced launch failures and in-orbit malfunctions. Meanwhile, the cost of the weather satellite program led to suggestions to cut the number of satellites and to sell them to private industry.

TIROS-N Polar Orbiters

The polar orbiters introduced in the late 1970s were of the new TIROS-N configuration. They replaced the ITOS, or TIROS-M configuration satellites. Again RCA was the prime contractor. RCA based the TIROS-N design on the Block 5D spacecraft design it had developed for the Defense Meteorological Satellites, already in orbit. TIROS-N looked nothing like TIROS-M. Its central structure, an elongated Equipment Support Module (ESM), with electronic equipment inside, carried antennas and instruments on its exterior. The cross-section of the ESM was a five-sided figure. One of the five sides always faced Earth and carried the antennas. The other four sides each had a row of three pinwheel louvers for thermal control. The Reaction Control Equipment Support Structure (RSS) on one end of the ESM housed a solid-fuel apogee kick motor that placed each spacecraft into the proper Sun-synchronous orbit ten minutes after launch. A boom connected a long, rectangular solar array to the RSS. In orbit, the solar array rotated about the boom to ensure exposure to sunlight. The other end of the ESM carried an instrument mounting platform, to hold the imaging and sounding instruments.[1] In its launch configuration (without the solar panel deployed), a TIROS-N spacecraft measured 148 inches in length and 75 inches in diameter. At launch it weighed 3,095 pounds but lost nearly half of that weight within a forty-five second period by expending the solid fuel of its apogee kick motor.[2]

TIROS-N satellites created day-and-night images with an Advanced Very High Resolution Radiometer (AVHRR), an improved version of the ITOS VHRR. The AVHRR could

Figure 70. An artist's conception of the TIROS-N satellite in orbit. (Photograph courtesy of NASA)

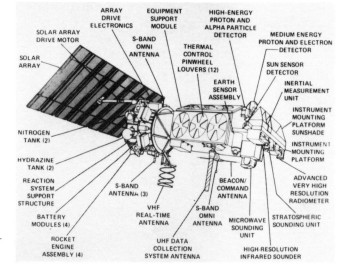

Figure 71. TIROS-N's major meteorological instruments were mounted on one end of the craft, at the right in this picture. (Photograph courtesy of NASA)

discern areas of cloud, ice, and snow cover, and it could give surface and cloud-top temperatures and pictures of atmospheric water vapor. At a resolution of ⅔ of a mile, the AVHRR could record pictures for transmission to CDA stations or read out pictures directly to specially equipped High Resolution Picture Transmission (HRPT) stations. At a reduced resolution of 2½ miles, pictures for entire orbits could be recorded, or pictures could be sent directly to standard APT stations.

The satellites carried a trio of sounding instruments, collectively called the TIROS-N Operational Vertical Sounder (TOVS). One of the three instruments was a Nimbus HIRS adapted for TIROS-N, abbreviated HIRS/2.[3] HIRS/2 used twenty channels to detect absorption of infrared radiation by several atmospheric components, so it could sound for temperature, water-vapor content, or ozone content. A HIRS/2 sounding covered an area on the Earth's surface 19 miles in diameter. To cover a larger surface area, the instrument would scan to take fifty-six soundings in a row perpendicular to the orbital path.[4] The United Kingdom's Meteorological Office provided the second TOVS instrument, the Stratospheric Sounding Unit (SSU), based on two radiometers already tested on Nimbus satellites. It used three infrared channels to provide temperature soundings of the stratosphere.[5] The SSU produced eight fields of view in a row 900 miles wide, perpendicular to the orbital path. The third TOVS instrument was a Microwave Sounding Unit (MSU) that could sound through clouds, unlike the two infrared sounders. The MSU sounded a circular area 70 miles in diameter and created eleven such soundings in an east–west row.[6]

An on-board data collection system called Argos could determine the position of platforms—either stationary surface platforms or free-floating balloons and buoys— and retrieve their meteorological or oceanographic data. The Centre National d'Etudes Spatiales (CNES), the French space agency, provided Argos instruments for the TIROS-N satellites and processed and disseminated Argos data.[7]

The first satellite of the series was launched in October 1978 and was named TIROS-N. NASA paid for this first satellite, which was to serve as the operational prototype for the program.[8] As with the ITOS satellites, NOAA wanted pairs of satellites in orbit simultaneously to acquire twice the coverage and ensure redundancy in case of a failure. TIROS-N orbited with NOAA 5 at first; the latter was replaced in June 1979 by NOAA 6, the second TIROS-N satellite. TIROS-N crossed the equator in a northward direction at approximately 3:00 P.M. local time. Its twin crossed

Figure 72. This image from TIROS-N shows the great Presidents' Day snowstorm that blanketed the eastern United States in 1979. (Photograph courtesy of NASA)

Figure 73. The same storm appears in upper central portion of this image from geosynchronous satellite SMS 1. (Photograph courtesy of NOAA)

Figure 74. The area of the storm on the following day, from TIROS-N. The coastline is clearly visible from New Jersey northward to Nova Scotia. (Photograph courtesy of NASA)

the equator in a southward direction at 7:30 A.M., local time.[9] The TIROS-N satellites as well as their successors, the Advanced TIROS-N, orbited at a 99° angle to the equator at an altitude of 500 miles.[10]

The GARP Global Weather Experiment

Along with the five geosynchronous satellites already in place, these two polar orbiters played an important role in the Global Weather Experiment. In 1978, after five years of planning and organizing, the experiment got underway. A preparatory year preceded the operational year of December 1978 through November 1979.[11] The experiment was truly a global effort, involving 140 countries that provided, collected, and studied data from ships,

aircraft, balloons, remote platforms, manned stations, and weather satellites. The timing of the experiment hinged on the availability of specific satellite systems, such as the five geosynchronous satellites stationed around the globe, and the TIROS-N polar orbiters. For the Global Weather Experiment, the TIROS-N and NOAA 6 AVHRR measured sea-surface temperatures and created cloud pictures; Argos packages collected data on sea-surface temperature (within one degree centigrade) and air pressure (within one millibar) from more than three hundred buoys drifting in the Southern Hemisphere oceans. The buoys had been donated by eight countries and deployed by ships from fourteen countries to take measurements in the normally data-sparse Southern Hemisphere. Using the same instruments, the satellites collected data from special balloons.[12] TOVS gave experimenters atmospheric sounding data. The ring of geosynchronous satellites provided images from which wind vectors (wind speed and direction) could be calculated. NOAA provided wind vectors from GOES data three times daily for the experiment.[13] In addition, geosynchronous satellites relayed weather data from collection packages installed on several wide-bodied jets for the ASDAR (Aircraft to Satellite Data Relay), another part of the experiment.[14]

At a 1986 workshop on the Global Weather Experiment, some of the results were discussed. Participants agreed that the experiment brought about important advances in understanding, analyzing, and forecasting weather conditions. Weather analyses generated from the experiment's data more accurately represented global weather than previous analyses. The data added two to three days to the time range for which advanced forecasts would be accurate. Satellites proved essential to the experiment. Valuable sounding data over the oceans could not have been collected in any other way. Data collected from buoys alone improved short-range forecasts in the Southern Hemisphere by twelve hours. The images from the geosynchronous satellites and the numerous wind vectors derived from those images fostered interest in maintaining a system of five geosynchronous satellites and in increasing studies of satellite-derived wind data.[15]

The next satellite in the TIROS-N series, NOAA-7, was launched in 1981. NOAA 7 was identical to its two predecessors, except that it included an additional AVHRR channel for better evaluation of sea-surface temperatures, and it carried an instrument to monitor contamination caused by mass accretion on the Earth-facing side of the ESM.[16]

The Advanced TIROS-N

Another similar spacecraft was built but put aside in favor of the first of the current series of polar orbiters, the Advanced TIROS-N (ATN). ATN is similar in appearance to TIROS-N, but the advanced version is 19 inches longer, and it carries additional instruments.[17] One of the primary reasons for jumping ahead to the ATN program was that the new satellites carried instruments involved in an international program of search and rescue called COSPAS/SARSAT, which involves the Soviet Union, France, Canada, and the United States. COSPAS is an acronym for the Russian *Kosmichesaia Sistem a Poiska Avariinykh Sudov* (Space System for the Rescue of Vehicles in Dis-

tress). The American part of the program is called SARSAT, for "Search and Rescue Satellite-Aided Tracking." The Soviet Union and the United States provide the satellites for the program, and France and Canada provide on-board instruments for the satellites. The satellites detect signals from standard emergency locator transmitters on downed aircraft or emergency position-indicating radio beacons from ocean vessels in distress. The satellites relay signals from a distressed craft to special ground stations, which process the signals to determine the location of the beacon and then alert rescue forces.

The first COSPAS-carrying satellite was launched in June 1982; NOAA 8 followed with its SARSAT payload in March 1983.[18] NOAA 9 and 10 also carried SARSAT payloads. The first satellite-aided rescue occurred in September 1982 when the COSPAS-equipped satellite located a Cessna 172 aircraft that had gone down in rugged mountains in British Columbia. The pilot and two passengers were saved. The American satellite's first rescue assist occurred in August 1983. Two Canadian men had been canoeing in the wilderness of northern Ontario, and their canoe overturned when they tried to pass some rapids. The two reached the shore, where they activated an emergency locator transmitter, whose signal was detected by NOAA 8. An airplane located the two men, and a helicopter rescued them shortly thereafter.[19] By 1986, the program had saved the lives of more than five hundred people and three dogs (from an Alaskan dogsled team). The problem has arisen, however, of numerous false alarms from beacons activated in places where no emergencies existed.

In addition to the SARSAT payload, ATN satellites carry two instruments to help determine the Earth's radiation budget, that is, the balance of incoming solar radiation with the radiation reflected, reradiated, and absorbed by the Earth. Any factor that upsets this balance, such as the greenhouse effect, could affect the weather. A Solar Backscatter Ultraviolet Spectral Radiometer (SBUV/2), a modified version of a Nimbus 7 instrument, measures ultraviolet radiation backscattered by the atmosphere.

A three-satellite project, the Earth Radiation Budget Experiment (ERBE), includes experiment packages carried by two ATN's and one carried by a satellite devoted solely to the project. Each of the three packages contains a scanning component and a nonscanning component. An ERBE scanner consists of three telescopes to collect radiation (1) reflected from the Earth, (2) emitted by the Earth and atmosphere, and (3) both reflected and emitted by the Earth. The nonscanner consists of five detectors. Four detectors monitor radiation from the Earth, and the fifth monitors energy from the Sun. The first ERBE package was carried by the Earth Radiation Budget Satellite (ERBS), deployed from the Space Shuttle Challenger in October 1984. NOAA 9 was launched shortly thereafter with the second package and NOAA 10 carried the third package into orbit in 1986. In preliminary studies, scientists have found ERBS and ERBE data quite valuable.[20]

The first ATN, NOAA 8, did not carry an ERBE package or an SBUV/2, but it carried dummies of those instruments to balance the satellite's weight. It carried the SARSAT payload, as well as the same instruments that TIROS-N satellites carried. NOAA 9 became the afternoon polar orbiter in December 1984, with a local, equator-crossing time of 2:20 P.M. northbound. NOAA 10 became the morning satellite, crossing the equator at 7:30 A.M. southbound. The launch of NOAA 10 on September 17, 1986, was the first successful launch of a United States civilian satellite after the Challenger explosion the previous January.[21] The identical NOAA 11 was launched in September 1988, to replace NOAA 9 as the afternoon satellite.[22]

Figure 75. An artist's conception of the Advanced TIROS-N (ATN) satellite in orbit. (Photograph courtesy of NASA)

Figure 76. RCA technicians prepare the NOAA F (9) satellite for shipment to Vandenberg Air Force Base in California for launch. (Photograph courtesy of GE Astro-Space Division)

Figure 77. Most of the ATN imaging and sounding instruments appear on the left in this picture of NOAA 9 before launch. The satellite's solar panels are stowed in the launch position. (Photograph courtesy of GE Astro-Space Division)

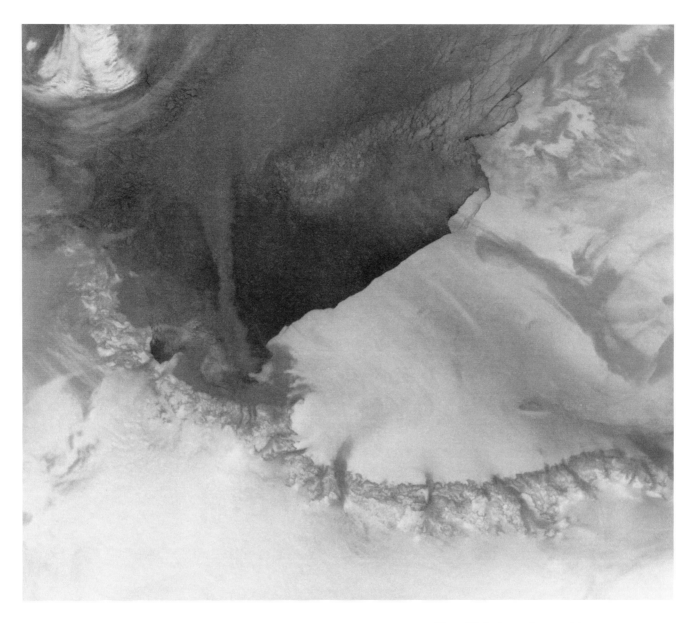

Figure 78. An image from an Advanced TIROS-N Advanced Very High Resolution Radiometer. (Photograph courtesy of NOAA)

Figure 79. A SARSAT satellite relays beacon distress signals from aircraft or vessels to a ground station, which pinpoints the distressed craft's location. The station relays this information to a rescue coordination center, which dispatches search and rescue personnel and equipment to the site of the emergency. (Photograph courtesy of NASA)

Figure 80. The Space Shuttle Challenger's remote manipulator deploys the Earth Radiation Budget Satellite (ERBS), on October 5, 1984. (Photograph courtesy of NASA)

The design and fabrication of these ATN/NOAA satellites changed hands in June 1986, when the General Electric Company bought RCA. RCA Astro Electronics, which created TIROS and all of its descendants, has become the GE Astro-Space Division.

New GOES Satellites

A new series of geosynchronous satellites complemented the polar orbiters. Hughes Aircraft won the contract for three new GOES. GOES 4, 5, and 6 generated images using the spin-scan method like their predecessors but used an improved VISSR imaging instrument that had a temperature and water-vapor sounding capability. NASA funded the new experimental instrument, called the VISSR Atmospheric Sounder (VAS), making GOES an experimental craft for soundings and an operational one for images. VAS can work as either an imager or a sounder but cannot be both simultaneously. VAS image resolution is .55 mile in the visible wavelengths and 4.27 miles in the infrared; its soundings have a maximum resolution of 18.6 miles over clear areas and 37 to 60 miles over partially cloud-covered areas.[23] Its twelve infrared channels help VAS to produce images, as well as data, on sea-surface temper-

atures, atmospheric temperatures, and the amount, distribution, and movement of water vapor.

The new GOES also employed a data collection system to relay weather data from remote platforms, and a Space Environment Monitor to measure solar activity and the intensity of the earth's magnetic field.[24]

NASA had originally hoped to launch GOES 4, 5, and 6 on the Space Shuttle, but because of delays in the Space Shuttle program, all were launched by Delta launch vehicles.[25] GOES 4 was launched on September 9, 1980, and was placed in geosynchronous orbit at 98° west. VAS took the first sounding from geosynchronous orbit one month later, after which NASA completed its usual checkout and turned over GOES 4 to NOAA. GOES 4 later replaced GOES 3 as the western satellite. GOES 5 was launched on May 22, 1981, and after its checkout period at 85° west, NOAA moved it to its operational eastern station at 75° west.[26] GOES 6 was launched in April 1983.

GOES 4, 5, and 6 were able to create images of water-vapor distribution as Europe's Meteosat did. The water-vapor images were so popular with meteorologists that NOAA distributed these images instead of the usual infrared images at 0530, 1130, 1730, and 2030 GMT. Water-vapor structures often appear in cloudless areas and can

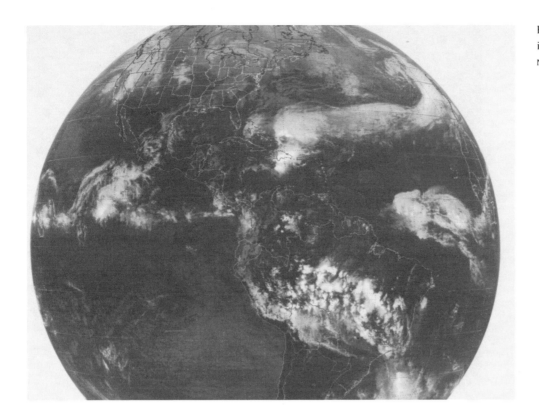

Figure 81. A recent GOES visible image. (Photograph courtesy of NOAA)

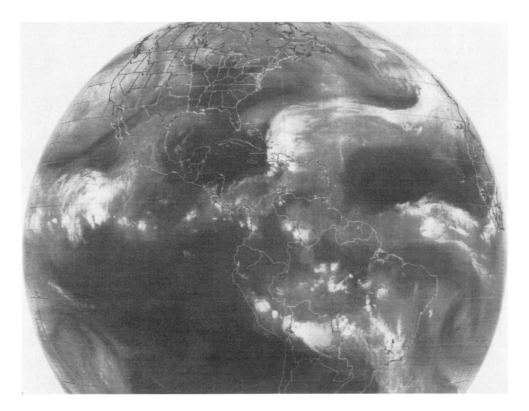

Figure 82. An image taken in the water-vapor channel at about the same time as figure 81. (Photograph courtesy of NOAA)

be observed and tracked from successive geosynchronous images, just as clouds can be tracked in visible images. Meteorologists at the University of Wisconsin began experimentally deriving wind vectors for the middle levels of the atmosphere by studying the vapor distributions on a MCIDAS system.[27]

The MCIDAS system was improved and expanded in the early 1980s. Originally built on minicomputers, MCIDAS was now based on mainframe computers. Several types of pictures, such as analysis maps, various satellite images, or wind-vector plots, could be superimposed on the display.[28] In addition, MCIDAS was used to eliminate unacceptable individual sounding samples from otherwise valid arrays of data from the polar orbiters.[29]

During the early to mid-1980s, the polar orbiter and geosynchronous programs suffered several launch vehicle and satellite instrumentation failures. NOAA B, a TIROS-N satellite scheduled to become NOAA 7, reached a useless, highly elliptical orbit because of an Atlas booster malfunction.[30] NOAA 8 was launched successfully, but malfunctions of its on-board clock sent the satellite tumbling time and again, for weeks at a time. The final malfunction of the clock caused an explosion of an over-charged battery, forcing the satellite into retirement.[31]

The GOES program also suffered from what television weathercasters aptly described as "a light bulb burning out." The light bulb was part of an encoder system that signaled a ratchet mechanism to move the spin-scanning mirror downwards on each scan to create a full picture. With a design life of five years, the two encoder lamps on the GOES 2 failed within one year and six months; those on GOES 3 failed within two years and nine months. In 1982, after two years' service, GOES 4 lost VAS capabilities due to an electronics problem unrelated to the encoder lamps. Yet in 1984, both encoder bulbs on GOES 5 failed within one week of each other, three years and three months after launch. Although GOES 4 and 5 had lost their imaging capabilities, they continued relaying data from ground stations. GOES 6, the only remaining United States geostationary weather satellite, was moved to 98° west so that it could cover both United States coasts. GOES 6 had been equipped with four, instead of two, encoder lamps. Following the lamp failures of GOES 2 and 3, Hughes, NOAA, and NASA tested similar lamps and found that fogging inside the bulbs may have caused the failures.[32] At NOAA's request, Hughes acceler-

Figure 83. GOES 6 created this image while it was the sole United States geosynchronous satellite, located between the normal east and west stations. (Photograph courtesy of NOAA)

Figure 84. The GOES-G spacecraft prior to rocket fairing installation. Its launch vehicle had to be de- stroyed shortly after launch, due to a failure. (Photograph courtesy of NASA)

ated production of GOES G and H, the next two in the series.[33] GOES G was not ready early but was on schedule for launch on May 3, 1986. NASA officials hoped that the very reliable Delta vehicle assigned to the GOES launch would accomplish the first successful launch since before the Challenger accident, but one minute into flight, a first-stage engine stopped, sending the Delta off course. A range safety officer destroyed the Delta and its $57.5 million payload.[34] Fortunately, GOES H was successfully launched on February 26, 1987.[35]

While NOAA was grappling with the problem of spacecraft failures, it simultaneously had to grapple with Reagan Administration proposals to cut its satellite budget. One proposal, to sell the satellites to private industry, originated at the Communications Satellite Corporation (COMSAT). An earlier budget-cutting proposal to commercialize the Landsat Earth-remote-sensing satellites drew interest from COMSAT, but the corporation wanted the weather satellites included in the deal. Top officials from the Office of Management and Budget (OMB) and the

Commerce Department supported the idea,[36] and President Reagan officially endorsed the transfer in March 1983.[37] Two months later, Deputy Commerce Secretary Guy Fiske resigned after Secretary Malcolm Baldridge admitted before the House Science and Technology Subcommittee that COMSAT had offered Fiske an executive position during the time that he had been exploring possibilities for the satellite sale.[38] Later that year, both the House and Senate passed nonbinding resolutions against the weather satellite sale to clearly demonstrate, according to Representative Don Fuqua, "that the Congress does not consider such a transfer to be in the national interest." In addition, the Senate voted to halt any Commerce funding that would be used to arrange the transfer.[39]

Meanwhile, OMB proposed cutting weather satellite costs in another way: by cutting the number of polar orbiters from two to one. This plan would cut in half the amount of polar-orbit data available to the 120 nations depending on it and would eliminate the guarantee of backup coverage in the event of failure. OMB argued that data from the Defense Meteorological Satellites could be used in an emergency; however, the defense satellites carry instruments that provide data that is inadequate for NOAA's mathematical forecasting models, and the defense satellites cannot send APT's. These concerns, centering on the potential downgrade of forecast quality, influenced Congress to vote down the proposal every year that it was introduced.[40]

At the time that OMB was making these proposals, NOAA sought help, in the form of funding or hardware, from other nations for maintaining the two-satellite system. When NOAA raised the issue at the 1984 Economic Summit of Industrialized Nations, participants decided to form an International Polar Orbiting Meteorological Satellite (IPOMS) group. IPOMS examined the problems and possibilities of international cooperation in the polar orbiter program and unanimously supported the need to maintain two satellites.[41] In 1986, IPOMS and the European Space Agency endorsed a system of one United States and one European polar orbiter, with a possible third platform mounted on future United States space stations.[42]

These plans, along with NOAA's plans for new polar orbiting and geosynchronous systems, will help to maintain the flow of satellite data that meteorologists use in their complicated analyses, that professionals and weather buffs collect on APT stations, and that television viewers see on the daily weather reports.

6

Tomorrow's Forecast

Over the last three decades, weather satellites have evolved into operational tools that give meteorologists many types of valuable data. Today, data from both the polar orbiters and the geosynchronous satellites are incorporated into the global mathematical weather models, and both types of satellites make their own specific contributions to the science of weather prediction.

For example, meteorologists map global sea-surface temperatures using data from polar orbiters. Sea-surface data have been particularly helpful in identifying El Niño, the phenomenon of unusual warming of the Pacific waters off the coast of South America, which affects weather on a large scale. El Niño would be much more difficult to detect by using ship reports alone. The polar orbiters are also particularly helpful to meteorologists by providing soundings over the oceans. Sounding data from the northern Pacific helps in predicting winter storms, and Atlantic sounding data helps meteorologists analyze those fast-developing winter storms that sometimes affect the eastern coast.

GOES satellite data give meteorologists the extremely important ability to monitor tropical storms, including hurricanes, from their original development stages through their final dissipation. No other method of data collection can provide for this constant monitoring; often, the developing stages of hurricanes occur out of range of meteorological aircraft.

Severe thunderstorms can be watched on GOES pictures before they appear on radar; water drops in the storms must reach a certain size before radar can detect them. By studying the shape and texture of clouds in GOES pictures, meteorologists can estimate the amount of rainfall that the clouds will produce. These rainfall estimates help local weather stations know when to warn the public about possible flash flooding. Soundings from GOES satellites help in studying unstable atmospheric conditions that may develop into severe weather.[1] Meteorologists will continue to require satellite data for use in mathematical models, analyses, and forecasts. New polar-orbiters and geosynchronous satellites will provide these data for years to come.

In the near future, six additional ATN satellites will be launched to maintain polar-orbit coverage. NOAA D, an older TIROS-N configuration craft, will also be launched. It had been put aside in favor of an ATN satellite that carried the first SARSAT payload.[2]

In the more distant future, a modified ATN satellite may be equipped with a sensor that can give direct information on winds. This sensor, called "lidar" for light detec-

Figure 85. The future GOES-NEXT spacecraft. GOES- I, J, and K will be of this configuration. In orbit, these spacecraft will employ three-axis stabilization to eliminate the need to spin. The imaging, sounding, and search and rescue (SAR) instruments all face out of one Earth-viewing side. A solar sail/boom will balance the pressure of the solar wind. (Photograph courtesy of Ford Aerospace and Communications Corporation)

tion and ranging, employs a laser to collect data actively rather than passively. That is, instead of carrying sensors that passively detect radiation coming from the Earth as the AVHRR and the current sounding instruments do, a lidar-equipped satellite could send its own laser radiation toward the Earth. The lidar sensor would use aerosol particles as wind tracers. The ways in which the laser light is scattered by suspended aerosol particles can tell something about the movement of the particles and thus the associated wind. Because of the potential for collect-

ing wind data, the proposed ATN satellite with lidar has been named the Windsat Free-Flyer. Lidar techniques for detecting temperature and humidity are also being studied, but as yet the use of lidar for meteorological application is still in the research-and-development stage.[3]

Aside from the continued employment of ATN satellites, a different polar-orbiting system may be established by NASA. As part of its space station program, NASA is hoping to orbit polar platforms that can be serviced and repaired by Space Shuttle astronauts. Unlike current satellites, a polar platform's lifetime would not necessarily end with the failure of one critical component. One platform being considered by NASA and NOAA is called EOS —the Earth Observing System. EOS would carry a number of meteorological instruments in addition to several geologic and geodesic instruments.[4]

A new generation of GOES satellites will ensure geosynchronous coverage well into the 1990s. Ford Aerospace won the contract to build five of the new craft, dubbed "GOES- NEXT." The new satellites will look nothing like previous GOES and will abandon spin stabilization and the spin-scan technique altogether. Each GOES- NEXT craft will be box shaped, with all meteorological sensors and communications equipment on one Earth-facing side. An on-board three-axis stabilization system will eliminate the need to spin the craft. Unlike the dual-function VAS on previous GOES, the imaging and sounding instruments on GOES-NEXT will be separate so that imaging and sounding can occur simultaneously. Furthermore, these GOES-NEXT sensors will have higher resolutions and more channels than the VAS. The first GOES-NEXT will probably be launched in the early 1990s.[5]

The future NOAA and GOES-NEXT satellites, and perhaps the polar platform, may play a role in a new global research project. The project has been promoted by NASA, the National Science Foundation (NSF), and the National Academy of Sciences. It would be similar to GARP, but instead of focusing on the atmosphere, the project would attempt to make a comprehensive study of the entire Earth, including air, sea, land, and living organisms.[6] NASA proposed the project under the name "Earth System Science." In 1986, the NASA Advisory Council made a report on Earth System Science that included recommendations to NASA, NOAA, and NSF. According to that report, the ambitious goal of the program is "to obtain a scientific understanding of the entire Earth System on a global scale by describing how its component parts and their interactions have evolved, how they function, and how

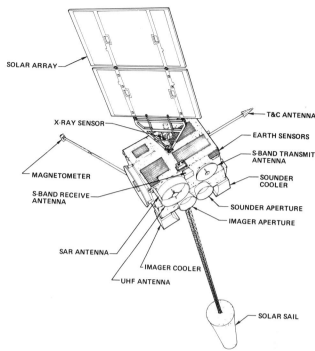

Figure 86. Diagrammatic depiction of the future GOES-NEXT spacecraft. (Photograph courtesy of Ford Aerospace and Communications Corporation)

they may be expected to continue to evolve on all time scales." The report also offers a challenge to the program, "to develop the capability to predict those changes that will occur in the next decade to century, both naturally and in response to human activity."[7]

A great deal of planning, funding, and interagency and international cooperation still lies ahead. But the idea of a global research program has incited much interest and enthusiasm in the scientific community and within the pertinent federal agencies.[8] Should the global program become a reality, weather satellites will doubtless play an essential role as data collectors and as the eyes that continually watch our weather.

7

Artifacts in the Collection of the National Air and Space Museum Relating to Rocket Photography and Weather Satellites

The Smithsonian Institution's National Air and Space Museum has in its collection several artifacts related to early photography from rockets and to the various weather satellite programs. This section lists some of the more significant of these artifacts.

The full-scale weather satellites in the museum's collection include models, engineering models, and prototypes used for testing, and flight-qualified spares that were not launched. For some programs, especially the more recent ones, the only spacecraft built were those that would definitely be launched, because of the high cost of developing, testing, and building each spacecraft. Therefore, for these programs the museum must use full- or smaller-scale models for adequate representation in its collection and displays.

Other than full-scale versions of the satellites, artifacts relating to the subjects of this book include a rocket camera, a photomosaic, a nephanalysis, and unlaunched satellite instruments.

Aerobee Camera and Photomosaic. As mentioned in chapter 1, the first high-altitude picture of a tropical storm was created from pieced-together pictures taken by a camera mounted inside an Aerobee rocket. The Naval Research Laboratory (NRL) launched this particular rocket, a developmental predecessor of the Aerobee-Hi rockets, from White Sands Proving Ground on October 5, 1954. The purpose of the flight was to test the new rocket's performance.[1] Two motion picture cameras would document the movements of the rocket. One of the cameras, now a museum artifact, happened to film a tropical storm as the rocket rolled about its axis. The rolling motion provided the means for strips of images to be created.[2] From 181 frames of the film, NRL scientist Otto Berg created images and pieced them together to make one large image—the photomosaic showing the tropical storm. For this achievement, Berg received the Navy's Superior Achievement Award.

On May 25, 1962, the NRL donated the original photomosaic and the camera with its protective flight case to the National Air Museum. At that time, so few space artifacts had been acquired by the museum that "Space" had not yet been incorporated into the museum's name, and a presentation ceremony accompanied the arrival of almost every new space artifact. The museum held a ceremony for NRL to officially present the photomosaic, camera, and case on October 5, 1962, the eighth anniversary of the Aerobee flight that provided

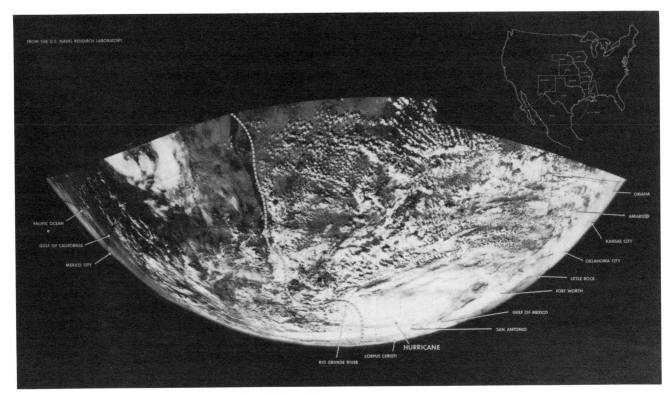

Figure 87. The Aerobee mosaic as   graph courtesy of the National Air
it was displayed at the National   and Space Museum)
Air and Space Museum. (Photo-

the pictures. Captain Arthur E. Krapf, NRL director, presented the artifacts; James Bradley, assistant secretary of the Smithsonian Institution, accepted them.

The photomosaic, camera, and case were displayed for many years in the Air and Space Building of the National Air Museum, and later in the National Air and Space Museum, which opened in 1976.

Three TIROS Artifacts.   The primary artifact representing the TIROS program is prototype T-2A. The T-2A performed numerous tests to help RCA and NASA engineers decide upon which improvements and adjustments to make on the flight models. Testing of the T-2A began in late June 1960 and continued for approximately four months. Tests included standard performance evaluation tests, noise vibration tests, and thermal vacuum tests, which involved placing the prototype into a vacuum chamber and exposing it to temperature swings from 0° to 60° centigrade.[3]

NASA presented the TIROS prototype to the National Air Museum at a ceremony on April 1, 1965, the fifth anniver-

Figure 88. Dr. Hugh L. Dryden,   the Smithsonian on April 1, 1965.
deputy administrator of NASA, pre-   (Photograph courtesy of the Na-
sents the TIROS T-2A prototype to   tional Air and Space Museum)

sary of the first TIROS launch. Speakers included high-level officials from the Smithsonian, NASA, the Weather Bureau, and the WMO. The prototype craft was immediately put on display in the Smithsonian's Arts and Industries Building, at that time part of the National Air Museum.

After the opening of the current National Air and Space Museum building, TIROS prototype T-2A resided first in the former "Satellites" gallery and then, beginning in 1985, in the "Looking at Earth" gallery.

In addition to the prototype spacecraft, the museum has in its collection a TIROS vidicon, similar to the tube inside each of the TIROS cameras. One end of the tube has the square, five-hundred-line picture-scanning area with tiny reference marks. The images of such reference marks on flight vidicons would appear in TIROS pictures.

The artifact was donated by RCA in 1976. It was displayed for a number of years in the museum's "Satellites" gallery.

The National Air and Space Museum's collection also includes a copy of a nephanalysis created from TIROS data. This particular nephanalysis was the first based on weather satellite data to be distributed to the meteorological community to use. It is autographed by the two meteorologists who prepared it, James B. Jones and J. H. Connover.

J. Gordon Vaeth of the National Weather Satellite Center personally donated the nephanalysis in February 1965.

Nimbus Artifacts. A full-scale model represents the Nimbus spacecraft in the collection. The General Electric Company's Missile and Space Division presented the model to the National Air and Space Museum on May 16, 1967, in a ceremony that commemorated one successsful year in space for Nimbus II.

Following its display in the old Air and Space Building,

Figure 89. The TIROS T-2A in its original display. (Photograph courtesy of the National Air and Space Museum)

Figure 90. TIROS T-2A on display in the "Looking at Earth" gallery. A wide-angle camera is at the bottom of the picture: a narrow-angle camera can be seen on the right side, to the upper left of the display mannequin. (Photograph courtesy of the National Air and Space Museum)

the model spent many years on display at the International Space Hall of Fame in Alamogordo, New Mexico, while on loan from the National Air and Space Museum.

More recently, the National Air and Space Museum acquired from NOAA three SIRS instruments similar to those carried on Nimbus craft. As mentioned in chapter 4, the actual SIRS-A was carried by Nimbus 3, and provided data for the first atmospheric sounding from space. SIRS-B, carried by Nimbus 4, was an improved version of SIRS-A.

The museum has prototype SIRS-A and SIRS-B instruments and a SIRS-B flight spare. For display purposes, the SIRS-A prototype had been partially cut away, and the SIRS-B prototype had been disassembled to expose its interior prior to NOAA's December 1986 donation.

In addition to imaging and sounding instruments, the Nimbus Program tested tracking and surface location instruments. The RAMS instrument carried by Nimbus 6 helped NASA personnel track Japanese explorer Naomi Uemura during his solo dogsled journey to the North Pole and his subsequent trek across Greenland, both between March and August 1978. He carried a data collection instrument that could transmit to Nimbus 6. The satellite relayed time, position, temperature, and barometric pressure data to NASA's GSFC, where Uemura's journeys were tracked.

In August 1978, Uemura donated his data collection platform, antenna, and battery pack to the National Air and Space Museum.

TOS Flight Spare. The TOS Program is represented in the museum's collection by a flight-qualified spare spacecraft. Designated TOS-H, this flight spare would have become ESSA 10 had it been launched. Like the even-numbered ESSA satellites, TOS-H has two APT cameras mounted in its sides.

In February 1975, NOAA donated the satellite to the museum for incorporation in the exhibition of meteorological satellites then being planned for the new (current) museum building. TOS-H was displayed with the TIROS T-2A prototype and the TIROS-M electrical test model for approximately eight years in that exhibition.

ATS 1 Engineering Test Model. The National Air and Space Museum has in its collection the full-scale Y-1 engineering test model of ATS 1. Unfortunately, this artifact, which represents the satellite that took the first geosynchronous weather picture, lacks its spin-scan cloud camera. Many other components remain intact,

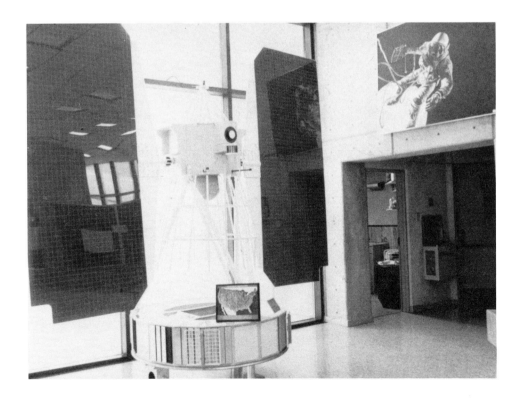

Figure 91. The Nimbus full-scale model, while on display at the Space Center museum in Alamogordo, New Mexico. (Photograph courtesy of the Space Center)

Figure 92. The TOS H satellite shortly after installation in the former "Satellites" gallery. A lens cap covers the camera. (Photograph courtesy of the National Air and Space Museum)

The first of these artifacts to be acquired was the TIROS-M electrical test model. (TIROS-M was the designation of ITOS 1 prior to launch). Like the TIROS-M that was launched, this artifact has a pair of vidicon cameras, a pair of APT cameras, and a pair of scanning radiometers. It has real solar cells and thermal blankets, and most internal components are present.

Like the TOS-H flight spare, the museum acquired the TIROS-M test model for display in the current building, which opened in 1976. RCA donated the artifact in October 1975. In 1985, it was placed in the museum's "Looking at Earth" gallery.

Only two flight-quality ITOS spacecraft were not launched; both are now in the collection of the National Air and Space Museum. One of these two craft was first built in the original ITOS configuration (with cameras) and designated ITOS- C. As early as 1970, NASA had considered retrofitting ITOS-C with two VHRR and two VTPR to create a second-configuration ITOS craft, to be launched after ITOS-G, if necessary. In July of that year, NASA and ESSA considered dropping this plan in favor of a plan to retrofit ITOS-C with just enough equipment to ensure continuous coverage, as well as some experimental instruments.[4]

In May 1971, NASA rejected this second plan for its semi-experimental spacecraft because of some flight difficulties with ITOS 1 and NOAA 1 and anticipated delays in the approval to begin the TIROS-N program. NASA decided to proceed with the original plan, to retrofit ITOS-C with VHRR and VTPR instruments and launch the craft some time after the launch of ITOS-G.[5]

Following its reconfiguration, ITOS-C was renamed ITOS E-2. Late in 1976, NASA tentatively planned to launch ITOS E-2 in June 1977, if NOAA 4 or 5 experienced some type of critical failure in orbit.[6] No failure occurred, so no additional ITOS craft were launched.

NOAA transferred ITOS E-2 to the National Air and Space Museum in 1978. It was displayed for several years in Japan and currently is on loan to the California Museum of Science and Industry.

The National Air and Space Museum also acquired the only other unflown, flight-quality craft of this series, ITOS I. Like NOAA 2 and subsequent satellites of the series, ITOS I was equipped with VHRR and VTPR instruments. NASA could have launched ITOS I if necessary.

The International Communication Agency owned ITOS I immediately prior to its 1980 transfer to the museum. In 1982, the museum loaned the satellite to the Musée de l'Air et de l'Espace in Paris.

however, including the UHF system, solar panels, and antennas.

The ATS Y-1 engineering test model was transferred to the museum from NASA's GSFC in May 1975.

ITOS Artifacts. As discussed in chapter 4, ITOS design underwent a transformation midway through the program. The original design, that of satellites ITOS 1 and NOAA 1, included vidicon cameras and APT cameras for daytime imaging and scanning radiometers for nighttime imaging. The reconfigured ITOS represented by NOAA 2 through 5, included no cameras, had VHRR for imaging, and had VTPR for added sounding capability.

ITOS are well represented in the NASM collection by three full-scale spacecraft.

Figure 93. The TIROS-M electrical test model on display in the "Looking at Earth" gallery. (Photograph courtesy of the National Air and Space Museum)

Figure 94. The ITOS E-2 flight spare sitting in the base of its shipping crate while in Japan. The solar panels are folded down. (Photograph courtesy of the National Air and Space Museum)

TIROS-N Model. A one-tenth scale model of TIROS-N represents the TIROS-N series of polar orbiters in the collection. The museum acquired the model for the "Looking at Earth" exhibition, which opened in 1985. RCA donated the model in August 1984.

GOES Full-Scale Model. The GOES program is represented by a full-scale model similar to GOES-G and H. GOES-G was destroyed because of a launch vehicle failure shortly after its launch on May 3, 1986. GOES-H was successfully launched on February 26, 1987.

Hughes Aircraft Company donated the model after having it built specially for the museum. The model was placed on display in the "Looking at Earth" gallery in 1985.

Figure 95. The TIROS-N model on display in the "Looking at Earth" gallery. (Photograph courtesy of the National Air and Space Museum)

Figure 96. The GOES full-scale model in the "Looking at Earth" gallery. (Photograph courtesy of the National Air and Space Museum)

Chronological List of Meteorological Satellites

Satellite	Launch Date	Remarks
Vanguard II	Feb. 17, 1959	Carried two photocell units
Explorer VI	Aug. 7, 1959	Carried image-scanning system
Explorer VII	Oct. 13, 1959	Provided data on Earth-Sun heat balance
TIROS I	Apr. 1, 1960	First successful cloud- imaging satellite
TIROS II	Nov. 23, 1960	Exceeded planned life span by nine months
TIROS III	Jul. 12, 1961	Discovered Hurricane Esther earlier than conventional methods could
TIROS IV	Feb. 8, 1962	
TIROS V	Jun. 19, 1962	
TIROS VI	Sep. 18, 1962	Provided meteorological support for two Mercury missions
TIROS VII	Jun. 19, 1963	
TIROS VIII	Dec. 21, 1963	Carried an APT camera
Nimbus 1	Aug. 28, 1964	Created the first nighttime cloud images
TIROS IX	Jan. 22, 1965	Wheel configuration; created images for first entire-Earth photomosaic
TIROS X	Jul. 1, 1965	First Weather Bureau-funded satellite
ESSA 1	Feb. 2, 1966	First fully operational weather satellite
ESSA 2	Feb. 28, 1966	Carried two APT cameras
Nimbus 2	May 15, 1966	
ESSA 3	Oct. 2, 1966	Carried two AVCS cameras
ATS 1	Dec. 7, 1966	Created first picture of Earth from geosynchronous orbit
ESSA 4	Jan. 1, 1967	Carried APT cameras
ESSA 5	Apr. 20, 1967	Carried AVCS cameras
ATS 3	Nov. 5, 1967	Created first color pictures of Earth from geosynchronous orbit
ESSA 6	Nov. 10, 1967	Carried APT cameras
Nimbus B	May 18, 1968	Failed to reach orbit
ESSA 7	Aug. 16, 1968	Carried AVCS cameras

Satellite	Launch Date	Remarks
ESSA 8	Dec. 15, 1968	Carried APT cameras; created a total of 265,136 images—many more than any other ESSA satellite created
ESSA 9	Feb. 26, 1969	Carried AVCS cameras
Nimbus 3	Apr. 14, 1969	First satellite with capability of providing atmospheric sounding data
ITOS 1	Jan. 23, 1970	Carried two AVCS cameras, two APT cameras, and two scanning radiometers
Nimbus 4	Apr. 8, 1970	
NOAA 1	Dec. 11, 1970	Craft similar to ITOS 1; first weather satellite funded by the newly formed NOAA
ITOS B	Oct. 21, 1970	Failed to reach orbit
NOAA 2	Oct. 15, 1972	First reconfigured ITOS series satellite (with no cameras, only radiometers); provided sounding data
Nimbus 5	Dec. 11, 1972	
ITOS E	Jul. 16, 1973	Failed to reach orbit
NOAA 3	Nov. 6, 1973	Reconfigured ITOS craft
SMS 1	May 17, 1974	First geosynchronous satellite dedicated solely to meteorological applications; assisted in GATE
NOAA 4	Nov. 15, 1974	Reconfigured ITOS craft
SMS 2	Feb. 6, 1975	
Nimbus 6	Jun. 12, 1975	
GOES 1	Oct. 16, 1975	First operational geosynchronous weather satellite
NOAA 5	Jul. 29, 1976	Reconfigured ITOS craft
GOES 2	Jun. 16, 1977	
GOES 3	Jun. 16, 1978	
TIROS N	Oct. 13, 1978	First TIROS Nconfiguration satellite
Nimbus 7	Oct. 24, 1978	Has provided data on the Antarctic ozone hole
NOAA 6	Jun. 27, 1979	TIROS N configuration
NOAA B	May 29, 1980	Failed to reach proper orbit

Satellite	Launch Date	Remarks
GOES 4	Sep. 9, 1980	First GOES with sounding capability
GOES 5	May 22, 1981	
NOAA 7	Jun. 23, 1981	TIROS N configuration
NOAA 8	Mar. 28, 1983	First satellite of ATN configuration; carried first SARSAT pay load
GOES 6	Apr. 28, 1983	
NOAA 9	Dec. 12, 1984	ATN configuration
GOES G	May 3, 1986	Destroyed by range safety officer shortly after launch
NOAA 10	Sep. 17, 1986	ATN configuration; carried ERBE
GOES 7	Feb. 26, 1987	
NOAA 11	Sep. 24, 1988	ATN

The Electromagnetic Spectrum

Wavelength (meters)	Type of Electromagnetic Radiation*	Data Provided by Weather Satellite Instruments Sensitive to These Wavelengths
10^{-12}	Gamma Rays	
10^{-11}		
10^{-10}		
10^{-9}	X Rays	
10^{-8}		
10^{-7}	Ultraviolet	Data on spatial and vertical distribution of ozone; data on solar UV energy.
10^{-6}	Visible	Cloud-cover images of the sunlit side of the Earth.
10^{-5}	Infrared	Cloud-cover images of both the sunlit and dark sides of the Earth; sounding data for partially cloudy and cloudless areas: information on the Earth's heat budget.
10^{-4}		
10^{-3}	Microwave	Sounding data (not dependent on a clear sky; rainfall and sea-ice mapping.
10^{-2}		
10^{-1}		
1	Radio	
10		
100		

*Divisions between the types of electromagnetic radiation are not in reality as well defined as they appear on this chart.

The historic Aerobee photomosaic taken on October 5, 1954, shows a tropical storm centered over Texas in the upper left corner.

ATS 1 created the first image of
the Earth from geosynchronous
orbit on December 9, 1966.
(Photograph courtesy of NASA)

Notes

Chapter 1

1. O.L. Godske et al., *Dynamic Meteorology and Weather Forecasting* (Boston: American Meteorological Society, 1957), 621–22.

2. Donald R. Whitnah, *A History of the United States Weather Bureau* (Urbana: University of Illinois Press, 1961), 12.

3. Patrick Hughes, *A Century of Weather Service: 1870 –1970* (New York: Gordon and Breach, Science Publishers, Inc., 1970), 16.

4. Friedman, Robert Marc, *Vilhelm Bjerknes and the Bergen School of Meteorology, 1918–1923* (Baltimore: Johns Hopkins University, doctoral dissertation, 1978), 1.

5. Whitnah, 191–92.

6. Godske et al., 625–28.

7. Godske et al., 642–43.

8. Godske et al., 646.

9. Willy Ley, *Rockets, Missiles, and Space Travel*, 2d rev. ed. (New York: The Viking Press, 1959), 254–89.

10. Lester Hubert and Otto Berg, "A Rocket Portrait of a Tropical Storm," *Monthly Weather Review* 83 (June 1955), 119 – 24.

11. Delmar L. Crowson, "Cloud Observations from Rockets," *Bulletin of the American Meteorological Society* 30 (Jan. 1949), 17–22.

12. Hubert and Berg, 119–24.

13. "Frontal Cloud System Pictures Obtained by Rocket," *Science* 129 (Jan. 1959), 198.

14. "Frontal Cloud System Pictures Obtained by Rocket," 199.

15. "Preliminary Design of an Experimental World-Circling Spaceship" (Douglas Aircraft Co. report no. SM-11827, Project Rand, May 2, 1946), 11–13; John H. Ashby, *A Preliminary History of the Evolution of the Tiros Weather Satellite Program* (NASA Goddard Space Flight Center, HHN-45, 1964), 8–13.

16. Eric Burgess, "The Establishment and Use of Artificial Satellites," *Aeronautics* 21 (Sept. 1949), 70–82.

17. Harry Wexler, "Observing the Weather from a Satellite Vehicle," *Journal of the British Interplanetary Society* 13 (Sept. 1954), 269–76.

18. S. Fred Singer, "Satellite Instrumentation—Results for the IGY," in *Ten Steps Into Space* (Philadelphia: Franklin Institute, 1958), 66.

19. W. G. Stroud and W. Nordberg, "Meteorological Measurements from a Satellite Vehicle," in *Scientific Uses of Earth Satellites*, ed. James A. Van Allen, 2d rev. ed. (Ann Arbor: University of Michigan Press, 1958), 119– 32.

Chapter 2

1. Stroud and Nordberg, "Meteorological Measurements from a Satellite Vehicle," 119–32; R. A. Hanel et al., "The Satellite Vanguard II: Cloud Cover Experiment," *IRE Transactions on Military Electronics* MIL-4 (Apr.–July 1960), 245–47.

2. *Project Able-3 Final Mission Report,* vol. 2 (Los Angeles: Space Technology Laboratories, 1960), 215–45.

3. *Juno II Summary Project Report,* vol. 1: *Explorer VII Satellite* (NASA Technical Note D-608, July 1961), 273 –90.

4. Ashby, *A Preliminary History of the Evolution of the Tiros Weather Satellite Program,* 8–13.

5. Jet Propulsion Laboratory Press Release (no date given).

6. H. I. Butler and S. Sternberg, "TIROS—the System and Its Evolution," *IRE Transactions on Military Electronics* MIL-4 (Apr. –July 1960), 248–56.

7. Ashby, 40–44.

8. NASA, *Final Report on the Tiros I Meteorological Satellite System,* 24, 40–44.

9. Richard Witkin, "U.S. Orbits Weather Satellite; It Televises Earth and Storms; New Era in Meteorology Seen," *New York Times* (Apr. 2, 1960), 1.

10. David Lawrence, " 'TIROS I' Is Called a Lesson to Critics of U.S. Program," *New York Herald Tribune* (Apr. 6, 1960).

11. John W. Finney, "U.S. Will Share TIROS I Pictures," *New York Times* (Apr. 5, 1960), 12.

12. NASA Press Release: #60-167, Apr. 8, 1960.

13. Jay S. Winston, "Use of TIROS Pictures in Current Synoptic Analysis," in *Proceedings of the International Meteorological Satellite Workshop* (Washington, D.C.: Government Printing Office, 1961), 95–96.

14. Ashby, 63.

15. D. G. King-Hele et al., *The R.A.E. Table of Earth Satellites, 1957–1986,* 3d rev. ed. New York: Stockton Press, 1987.

16. Ashby, 37–39.

17. William Nordberg, "Physical Measurements and Data Processing," in *Proceedings of the International Meteorological Satellite Workshop* (Washington D.C.: Government Printing Office, 1961), 118.

18. J. Gordon Vaeth, *Weather Eyes in the Sky* (New York: The Ronald Press Company, 1965), 28–31.

19. Morris Neiburger and Harry Wexler, "Weather Satellites," *Scientific American* 205 (July 1961), 81.

20. Ashby, 67–69.

21. Robert M. White, "Weather Satellite System," *The Military Engineer* 58 (July–Aug. 1966), 235.

22. Robert W. Popham, "A Personal Eye in Space," *Weatherwise* 37 (Apr. 1984), 77–78.

23. Richard LeRoy Chapman, *A Case Study of the U.S. Weather Satellite Program: The Interaction of Science and Politics* (Syracuse University, doctoral dissertation, 1967), 75–79.

24. Chapman, 90–99.

25. Chapman, 103–28.
26. Chapman, 136–38.
27. Chapman, 159–62.
28. Chapman, 185.
29. Chapman, 232–40.
30. Chapman, 244–46.
31. Chapman, 268.

Chapter 3

1. William K. Widger, Jr., *Meteorological Satellites* (New York: Holt, Rinehart and Winston, Inc., 1966), 153–56.

2. NASA News Release (NR# 61-18, Feb. 3, 1961).

3. Widger, 157.

4. Lester F. Hubert and Paul E. Lehr, *Weather Satellites* (Waltham, Mass.: Blaisdell Publishing Company, 1967), 36.

5. Hubert and Lehr, 44.

6. Rudolf Stampfl and Harry Press, "Nimbus Spacecraft System," *Aerospace Engineering* 21 (July 1962), 16–28.

7. Widger, 183–85.

8. William Nordberg, "Geophysical Observations from Nimbus I," *Science* 150 (Oct. 29, 1965), 559.

9. Harry Press and Wilbur B. Huston, "Nimbus: A Project Report," *Astronautics and Aeronautics* 6 (Mar. 1968), 57–58.

10. E. Raschke and W. R. Bandeen, "The Radiation Balance of the Planet Earth from Radiation Measurements of the Satellite Nimbus II," *Journal of Applied Meteorology* 9 (Apr. 1970), 215–38.

11. Nordberg, 564–68.

12. U.S. Department of the Interior/Geological Survey, *Infrared Techniques Help Prove Surtsey's Fiery Mantle* (Washington, D.C.: U.S. Government Printing Office, 1967).

13. U.S. Department of Commerce/ESSA, *ESSA Science and Engineering* (Washington, D.C.: U.S. Government Printing Office, 1968), 1.

14. D. G. King-Hele et al.

15. U.S. Department of Commerce/National Environmental Satellite Center, *Satellite Activities of the Environmental Satellite Services Administration* (Washington, D.C.: U.S. Department of Commerce, 1966), 2–5.

16. U.S. Department of Commerce and NASA News Release, "Fifth ESSA Satellite Launch Scheduled" (# ES67-39, Apr. 16, 1967).

17. Hughes, 149.

18. William J. Normyle, "Soviet Weather Satellite Photos Sent to U.S.," *Aviation Week and Space Technology* 85 (Sept. 26, 1966), 26–27.

19. NASA News Release, "APT Photos Help Save Mexican Cities" (NR# 68-203, Nov. 28, 1968).

20. *Washington Post,* Feb. 20, 1969, A10.

21. Widger, 216.

22. Morris Tepper, "A Solution In Search of a Problem," *Bulletin of the American Meteorological Society* 44 (Sept. 1963), 547.

23. G. P. Tennyson, "An Informal Discussion—Scientific Objectives of the TIROS K Mission" (Draft of a NASA report, Feb. 16, 1965).

24. Barry Miller, "Satellites Will Test Advanced Avionics," *Aviation Week and Space Technology* 81 (July 20, 1964), 60–65.

25. NASA News: Background Briefing on Applications Technology Satellite (1965).

26. NASA Press Kit, ATS-F (May 21, 1974), 11–12.

27. NASA News Release, "Millimeter Wave Experiment" (Oct. 25, 1970).

28. NASA Press Kit, ATS-B (Dec. 2, 1966), 24.

29. NASA News Release, "Alaskan Rescue Via ATS-I" (Oct. 25, 1970).

30. NASA Memorandum to H. J. Goett, director, and Dr. Townsend of Goddard Space Flight Center, from John F. Clark, associate administrator for Space Science and Applications, July 16, 1965.

31. NASA Press Kit, ATS-B (Dec. 2, 1966), 9.

32. NASA Press Kit, ATS-B (Dec. 2, 1966), 10.

33. NASA News Release, "Satellite Tracking" (NR# 68-111, July 26, 1968).

34. Hughes News Release, "Satellite Kisses Blarney Stone: Is Blessed with 'Gift of Tongues'" (Oct. 17, 1968); NASA News Release, "Apollo 7 Via ATS-III" (NR# 68-179, Oct. 14, 1968).

35. Michael L. Garbacz, NASA Memorandum for the Record, "ATS-1 Spin-Scan Camera Experiment Review" (Apr. 28, 1967).

36. Tetsuya T. Fujita et al., "Formation and Structure of Equatorial Anticyclones Caused by Large-scale Cross-equatorial Flows Determined by ATS-1 Photographs," *Journal of Applied Meteorology* 8 (Aug. 1969), 649–67.

37. D. N. Sikdar and V. E. Suomi, "On the Remote Sensing of Mesoscale Tropical Convection Intensity from a Geostationary Satellite," *Journal of Applied Meteorology* 11 (Feb. 1972), 37–43.

38. U.S. Department of Commerce/ESSA News Release, "ATS Spacecraft Set to Photograph Tornado Weather in ESSA/NASA Experiment" (Mar. 14, 1968).

39. U.S. Department of Commerce/ESSA News Release, "Continuous Satellite Photographs Now Available to Tornado Forecasters" (#ES 70-31, May 5, 1970).

40. U.S. Department of Commerce, *First Five Years of the Environmental Satellite Program–An Assessment* (Washington, D.C.: U.S. Department of Commerce, 1971), 2–3.

Chapter 4

1. King-Hele et al.

2. NASA News Release, "Nimbus Failure Report" (NR# 68-171, Oct. 4, 1968).

3. Kendrick Frazier, "Sounding From Above," *Science News*, 96 (Nov. 29, 1969), 509–11.

4. NASA/Goddard Space Flight Center Press Kit, "Nimbus E" (Dec. 1972), 6.

5. NASA Press Kit, "Nimbus F" (June 8, 1975), 12.

6. NASA Press Kit, "Nimbus F" (Sept. 8, 1978), 14–18.

7. Warren C. Wetmore, "Nimbus 3 Soundings Exceed Expectations," *Aviation Week and Space Technology* 90 (May 5, 1969), 32–33.

8. NASA News Release, "Nimbus Failure Report."

9. Gerald L. Wick, "Nimbus Weather Satellites: Remote Sounding of the Atmosphere," *Science* 172 (June 18, 1971), 1222–23.

10. D. H. Staelin et al., "Microwave Spectrometer on the Nimbus 5 Satellite: Meteorological and Geophysical Data," *Science* 182 (Dec. 28, 1973), 1339–41.

11. M. Steven Tracton and Ronald D. McPherson, "On the Impact of Radiometric Sounding Data upon Operational Numerical Weather Prediction at NMC," *Bulletin of the American Meteorological Society* 58 (Nov. 1977), 1201–09.

12. M. Ghil, M. Halem, and R. Atlas, "Time-Continuous Assimilation of Remote-Sounding Data and Its Effect on Weather Forecasting," *Monthly Weather Review* 107 (Feb. 1979), 140–65.

13. Joseph Steranka, Lewis J. Allison, and Vincent V. Salomonson, "Applications of Nimbus 4 THIR 6.7 Micrometer Observations to Regional and Global Moisture and Wind Field Analyses," *Journal of Applied Meteorology* 12 (Mar. 1973), 386–95.

14. NASA/Goddard Space Flight Center Press Kit, "Nimbus E" (Dec. 1972), 7.

15. Paul Julian, "Tropical Wind, Energy Conservation, and Reference Level," in *Nimbus 6 Random Access Measurement System Applications Experiments*, NASA SF-457 (Washington, D.C.: U.S. Government Printing Office, 1982), 8.

16. Lee Houzpins. "Dogsled Tracking," in *Nimbus 6 Random Access Measurement System Applications Experiments*, NASA SP-457 (Washington, D.C.: U.S. Government Printing Office, 1982), 79.

17. NASA Press Kit, "Nimbus F" (Sept. 8, 1978), 4–6.

18. NASA Press Kit, "TIROS-M" [Release # 70-2, Jan. 13, 1970].

19. King-Hele et al.

20. NASA/Goddard Space Flight Center, *ITOS* (Washington, D.C.: U.S. Government Printing Office, 1970), 5.

21. NASA News Release, "Second Weather Satellite Series Turnover" (NR# 70–96, June 15, 1970).

22. NASA/Goddard Space Flight Center, *ITOS*, 3.

23. The Commission on Marine Science, Engineering, and Resources first proposed the establishment of NOAA in January 1969, in its report, "Our Nation and the Sea." NOAA was to be an independent agency composed of ESSA, a number of marine programs, and the Coast Guard. Bryce Nelson, "Marine Com-

mission Invokes NOAA, Urges Refitting of Nation's Ark," *Science* 163 (Jan. 17, 1969), 263–65.

President Nixon announced the formation of NOAA in July 1970, but the new agency did not include the Coast Guard, and it was not independent; unlike ESSA, NOAA would be part of the Commerce Department. Nonetheless, the oceanographic community was generally satisfied with the new arrangement ("Getting Ocean Sciences Together," *Science News* 98 [July 25, 1970], 59.

24. *The National Oceanic and Atmospheric Administration*, U.S. Department of Commerce pamphlet (NOAA/ PA 84004, June 1984).

25. George H. Ludwig, "The NOAA Operational Environmental Satellite System—Status and Plans," unpublished paper in National Air and Space Museum accession files.

26. Ludwig.

27. NASA Press Kit, "ATS-F" (May 21, 1974), 50.

28. *An Introduction to GARP* (World Meteorological Organization and International Council of Scientific Unions, 1969), 15–18.

29. Ralph Corbell, Cornelius Callahan, and William J. Kotsch, eds., *The GOES/SMS User's Guide* (Washington, D.C.: NASA/Goddard Space Flight Center and the National Oceanic and Atmospheric Administration, 1976), 1.

30. "NASA Ordering Two Additional GOES Metsats from WDL," *Defense/Space Daily* (Oct. 24, 1973), 270.

31. NASA Press Kit, "SMS-A" (May 8, 1974).

32. NASA Mission Operation Report, Synchronous Meteorological Satellite B, #E-608-75-02 (Washington, D.C.: U.S. Government Printing Office, 1975).

33. Corbell, Callahan, and Kotsch, 7.

34. NASA Mission Operation Report, Synchronous Meteorological Satellite B, #E-608-75-02.

35. "A New Breed of Weather Satellite: The Fixed Stare," *Science News* 105 (May 25, 1974), 332–33.

36. "Shift in Scene," *Science News* 97 (Apr. 4, 1970), 342.

37. Craig Covault, "Satellites Guide Aircraft in Tropical Weather Study," *Aviation Week and Space Technology* 100 (July 1, 1974), 47–50.

38. NASA News Release, "First Operational Geosynchronous Weather Satellite Set for Launch" (NR# 75-270, Oct. 12, 1975).

39. NASA News Release, "Daily TV Weather Picture Coming from New Satellite" (NR# 75-10, Jan. 14, 1975).

40. Corbell, Callahan, and Kotsch, 1.

41. U.S. Department of Commerce News Release, "NOAA Environmental Satellite GOES Operational over Brazil" (NR# NOAA 76-1, Jan. 8, 1976).

42. NASA Mission Operation Report, GOES B (Washington, D.C.: U.S. Government Printing Office, 1977).

43. "Geostationary Operational Environmental Satellite (GOES 3) Launched Successfully June 16 from Cape Canaveral," *Aviation Week and Space Technology* 109 (July 3, 1978), 11.

44. Eric A. Smith, "The MCIDAS System," *IEEE Transactions on Geoscience Electronics* 13 (July 1975), 123; Gary C. Chatters and Verner E. Suomi, "The Applications of MCIDAS," *IEEE Transactions on Geoscience Electronics* 13 (July 1975), 137.

45. David Suchman and David V. Martin, "Wind Sets from SMS Images: An Assessment of Quality for GATE," *Journal of Applied Meteorology* 15 (Dec. 1976), 1265– 78.

46. Chatters and Suomi, 137.

47. John A. Ernst, "A Different Perspective Reveals Air Pollution," *Weatherwise* 28 (Oct. 1975), 215–16.

48. W. A. Lyons, J. C. Dooley, and K. T. Whitney, "Satellite Detection of Long-Range Pollution Transport and Sulfate Aerosol Hazes," *Atmospheric Environment* 12 (1978), 621–31.

49. Robert A. Sutherland et al., "A Real-Time Satellite Data Acquisition, Analysis and Display System—A Practical Application of the GOES Network," *Journal of Applied Meteorology* 18 (Mar. 1979), 355–60.

50. NASA Press Kit, "GOES C" (June 11, 1978).

51. ESA News Release (Nov. 14, 1978).

Chapter 5

1. NASA Mission Operation Report, NOAA-A, #E-615-79-01 (Washington, D.C.: National Aeronautics and Space Administration, 1979), 3–7.

2. NASA Press Kit, "TIROS-N" (Sept. 5, 1978).

3. NASA Mission Operation Report, NOAA-A, #E-615-79-01, 4–10.

4. W. L. Smith et al., "The TIROS-N Operational Vertical Sounder," *Bulletin of the American Meteorology Society* 60 (Oct. 1979), 1177–87.

5. NASA Mission Operation Report, NOAA-A, #E-615-79-01, 10.

6. Smith et al., 1178–82.

7. NASA Mission Operation Report, TIROS-N, #E-614-78-01 (Washington, D.C.: National Aeronautics and Space Administration, 1978), 3.

8. NASA Mission Operation Report, TIROS-N, #E-614-78-01, 1 –3.

9. National Environmental Satellite Service, *Satellite Activities of NOAA 1978* (Washington, D.C.: U.S. Department of Commerce, 1979).

10. King-Hele et al.

11. Robert N. Green, "Global Weather Experiment Data from the National Environmental Satellite Service," *Bulletin of the American Meteorology Society* 61 (Oct. 1980), 1225.

12. R. J. Fleming, T. M. Kaneshige, and W. E. McGovern, "The Global Weather Experiment: I. The Observational Phase Through the First Special Observing Period," *Bulletin of the American Meteorology Society* 60 (June 1979), 649–59.

13. Green, 1225.

14. James K. Sparkman, Jr., James Giraytys, and George J. Smidt, "ASDAR: A FGGE Real-Time Data Collection System," *Bulle-*

tin of the American Meteorology Society 62 (Mar. 1981), 394–400.

15. Donald R. Johnson, "Summary of the Proceedings of the First National Workshop on the Global Weather Experiment," *Bulletin of the American Meteorology Society* 67 (Sept. 1986), 1135–43.

16. NOAA/NASA Press Kit, "NOAA- C" (June 12, 1981), 2–7.

17. "TIROS-N Performance Gains Postpone Replacement Plan," *Aviation Week and Space Technology* 113 (July 21, 1980), 51.

18. NASA/Goddard Space Flight Center Fact Sheet, "Search and Rescue Satellite/COSPAS Program Description."

19. NASA News Release, "Two Canadians Rescued in First Assist by American Satellite" (NR# 83-129, Aug. 18, 1983).

20. ERBE Science Team, "First Data From the Earth Radiation Budget Experiment (ERBE)," *Bulletin of the American Meteorology Society* 67 (July 1986), 818–24.

21. "NOAA Weather Satellite Launched on Atlas E," *Aviation Week and Space Technology* 125 (Sept. 22, 1986), 51.

22. NASA News Release, "NOAA-H Weather Satellite To Be Launched' (NR# 88-126, Sept. 14, 1988).

23. Craig Covault, "New Satellite to Sample Data on Weather by Altitude," *Aviation Week and Space Technology* 113 (Sept. 15, 1980), 24–25.

24. NOAA/NASA Press Kit, "GOES E" (Apr. 27, 1981).

25. "GOES Going?" *Aerospace Daily* 98 (July 9, 1979); NASA Memorandum to Anthony J. Calio, associate administrator for Space and Terrestrial Applications, from John F. Yardley, associate administrator for Space Transportation Systems (Nov. 6, 1978); Telegram from Chester M. Lee, director, STS Operations, to Johnson and Kennedy Space Centers (July 9, 1979); Telegram from Chester M. Lee, director, STS Operations, to Johnson and Kennedy Space Centers and Goddard Space Flight Center (Sept. 19, 1979).

26. NASA Memorandum to James Beggs, administrator, from Burton I. Edelson, associate administrator for Space Science and Applications (July 23, 1985).

27. Tod R. Stewart, Christopher M. Hayden, and William L. Smith, "A Note on Water-Vapor Wind Tracking Using VAS Data on MCIDAS." *Bulletin of the American Meteorology Society* 66 (Sept. 1985), 1111–15.

28. V. E. Suomi et al., "MCIDAS III: A Modern Interactive Data Access and Analysis System," *Journal of Climate and Applied Meteorology* 22 (May 1983), 766–78.

29. Smith et al., 1182.

30. NASA News Release, "NOAA-B Environmental Monitoring Satellite Mission Unsuccessful" (NR# 80-82, June 3, 1980).

31. "NOAA Turns Off Satellite following Malfunction," *Aviation Week and Space Technology* 124 (Jan. 13, 1986), 21.

32. Jay C. Lowndes, "Another Lamp Failure Blinds GOES 5," *Aviation Week and Space Technology* 121 (Aug. 6, 1984), 22–24. NOAA reduced incentive payments to Hughes after the GOES 4 failure, and the contractor returned $3.8 million of performance payments to NOAA following the GOES 5 failure.

33. "National Oceanic and Atmospheric Administration Will Reduce Incentive Payments to Hughes following Loss of Weather Imaging Capability," *Aviation Week and Space Technology* 117 (Dec. 13, 1982), 13.

34. Edward H. Kolcum, "Delta Engine Shutdown Investigation Centers on Electrical Relay Box," *Aviation Week and Space Technology* 124 (May 12, 1986), 20–22.

35. Bruce D. Nordwall, "NASA Solves Problems, Corrects GOES-H Orbit," *Aviation Week and Space Technology* 126 (Mar. 9, 1987), 266–67.

36. M. Mitchell Waldrop, "A Silver Lining For Weather Satellites?" *Science* 226 (Dec. 14, 1984), 1289.

37. "Reagan Orders Satellite Sale," *Aviation Week and Space Technology* 118 (Mar. 14, 1983), 263.

38. M. Mitchell Waldrop, "Commerce Deputy Resigns over Satellite Sale," *Science* 220 (May 27, 1983), 934–35.

39. "House Rejects Sale of Satellites to Industry," *Aviation Week and Space Technology* 119 (Nov. 21, 1983), 18.

40. Waldrop, "Silver Lining," 1289.

41. "Dual Polar Satellites Draw International Support," *Aviation Week and Space Technology* 121 (Dec. 10, 1984), 27.

42. "Weather Satellite Operators Push for Two Polar Orbitors," *Aviation Week and Space Technology* 124 (Jan. 13, 1986), 149.

Chapter 6

1. Interview with Ron Gird, The Analytic Sciences Corporation, Arlington, Va., Feb. 27, 1987.

2. Interview with Laura Eberle, RCA Astro Electronics, Princeton, N.J., Feb. 24, 1987.

3. A. F. Bogdan et al., "Laser Remote Sensors for Space Applications," *RCA Engineer* 31 (May–June 1986), 59–69.

4. *Earth System Science: A Program for Global Change* (Washington, D.C.: National Aeronautics and Space Administration, 1986), 38–39.

5. "New GOES to Sharpen Severe Weather Tracking," *Aviation Week and Space Technology* 123 (Dec. 23, 1985), 56–57.

6. M. Mitchell Waldrop, "Washington Embraces Global Earth Sciences," *Science* 233 (Sept. 5, 1986), 1040–42.

7. *Earth System Science: A Program for Global Change*, 38–39.

8. Waldrop, "Washington Embraces," 1040–42.

Chapter 7

1. Homer E. Newell, *Sounding Rockets* (New York: McGraw-Hill Book Company, Inc., 1959), 68–70.

2. L. F. Hubert, and Otto Berg, "A Rocket Portrait of a Tropical Storm," *Monthly Weather Review* 83 (June 1955), 119–24.

3. TIROS II Meteorological Satellite System, Final Engineering Report (Astro Electronics Division, RCA Doc. #AED-582, Dec. 29, 1961), II, IV-23–IV-25.

4. Michael Garbacz, NASA Memorandum to the deputy associate administrator for Space Science and Applications (Applications), "Proposal to Utilize ITOS-C for Piggyback NASA Experiments," July 1, 1970.

5. John M. DeNoyer, NASA Memorandum to the deputy associate administrator for Space Science and Applications (Applications), "ITOS-C and its Relevance to the SATS program," May 25, 1971.

6. "NASA Plans Nineteen Space Flights in 1977," *Aviation Week and Space Technology* 105 (Dec. 13, 1976), 26.

All other information in this section comes from National Air and Space Museum presentation and accession files.

Suggestions for Further Reading

Butler, H. I., and S. Sternberg. "TIROS–The System and Its Evolution." *IRE Transactions on Military Electonics* MIL-4 (Apr.–July 1960): 248–56.

Chapman, Richard LeRoy. *A Case Study of the U.S. Weather Satellite Program: The Interaction of Science and Politics.* Syracuse University, doctoral dissertation, 1967.

Chetty, P. R. K. *Satellite Technology and Its Applications.* Blue Ridge Summit, Pa: Tab Books Inc., 1988.

Cowen, Robert G. "Weather Satellites Fulfilling the Promises." *Weatherwise* 37 (Apr. 1984): 64–67.

Emiliani, Guido, and Marciano Righini. "An S-Band Receiving System for Weather Satellites." *QST* 64 (Aug. 1980): 28–33.

Emiliani, Guido, and Marciano Righini. "Printing Pictures from 'Your' Weather Geostationary Satellite." *QST* 65 (Apr. 1981): 20–25.

Hanel, R. A., et al., "The Satellite Vanguard II: Cloud Cover Experiment." *IRE Transactions on Military Electronics* MIL-4 (Apr.–July 1960).

Henderson-Sellers, Ann, ed. *Satellite Sensing of a Cloudy Atmosphere: Observing the Third Planet.* London: Taylor and Francis Ltd., 1984.

Houghton, J. T., F. W. Taylor, and C. D. Rodgers. *Remote Sounding of Atmospheres.* Cambridge: Cambridge University Press, 1984.

Hubert, Lester F., and Paul E. Lehr. *Weather Satellites.* Waltham, Mass.: Blaisdell Publishing Company, 1967.

Hughes, Patrick. A Century of Weather Service: 1870–1970. New York: Gordon and Breach, Science Publishers, Inc., 1970.

———. "Weather Satellites Come of Age." *Weatherwise* 37 (Apr. 1984): 68–75.

Lehrain, David, and Michael Lightfoot. "Direct Satellite Readout." *Weatherwise* 38 (June 1985): 159–61.

Ley, Willy. *Rockets, Missiles, and Space Travel.* 2d rev. ed. New York: The Viking Press, 1959.

NASA. *Earth System Science: A Program for Global Change.* Washington, D.C.: 1986. 38–39.

NASA/Goddard Space Flight Center. *Final Report on the Tiros I Meteorological Satellite System.* Washington, D.C.: 1962 (Technical Report R-131).

Popham, Robert W. "A Personal Eye in Space." *Weatherwise* 37 (Apr. 1984): 76–82.

Stroud, W. G., and W. Nordberg. "Meteorological Measurements from a Satellite Vehicle," in *Scientific Uses of Earth Satellites,* ed. James A. Van Allen. 2d rev. ed. Ann Arbor: University of Michigan Press, 1958.

Vaeth, J. Gordon. *Weather Eyes in the Sky*. New York: The Ronald Press Company, 1965.

Waldrop, M. Mitchell. "Commerce Deputy Resigns Over Satellite Sale." *Science* 220 (May 27, 1983): 934–35.

———. "A Silver Lining For Weather Satellites?" *Science* 226 (Dec. 14, 1984): 1289–91.

———. "Washington Embraces Global Earth Sciences." *Science* 223 (Sept. 5, 1986): 1040–42.

Wexler, Harry. "Observing the Weather from a Satellite Vehicle." *Journal of the British Interplanetary Society* 13 (Sept. 1954): 269–76.

Whitnah, Donald R. *A History of the United States Weather Bureau*. Urbana: University of Illinois Press, 1961.

Widger, William K., Jr. *Meteorological Satellites*. New York: Holt, Rinehart and Winston, Inc., 1966.

Index